T0296390

New Methuselahs

Basic Bioethics
Arthur Caplan, editor

A complete list of the books in the Basic Bioethics series appears at the back of this book.

New Methuselahs

The Ethics of Life Extension

John K. Davis

The MIT Press
Cambridge, Massachusetts
London, England

This book was set in Stone Serif Std by Scribe Inc.

Library of Congress Cataloging-in-Publication Data is available.

ISBN: 978-0-262-03813-3 (hardcover)
ISBN: 978-0-262-55156-4 (paperback)

Contents

Contents

Introduction

This book is about the ethics of life extension. By "life extension," I don't mean end-of-life care for the elderly. I mean slowing or halting human aging. At its extreme, this would be endless youth, the sort of thing we know from myths and science fiction. Even if we didn't halt aging but merely slowed it down, we might look and feel youthful much longer than we do now—perhaps decades longer. Perhaps more.

Unlike most books about ethics and biotechnology, this book is about a technology that does not yet exist, and that requires an explanation. Why should we think life extension is plausible? Moreover, why would anyone think there is an ethical problem with keeping people youthful for longer? And why write about this now? Why not wait until life extension arrives and we know more about it?

To show that life extension is plausible, I must spend more time talking about science than is usual in an ethics book. (Most of that material appears in chapter 1; the rest appears in appendix A.) As wild as it may sound, many scientists now take life extension seriously. Geroscience—the branch of biology that studies aging—has made dramatic strides over the last 20 years. Researchers have already slowed aging in several species—including yeast, mice, and fruit flies—and learned that humans share many aging-related genes with those species. Many mainstream geroscientists now publicly declare that we may soon learn how to slow aging in humans. Some of them have even founded pharmaceutical companies to develop drugs for this. There is a rising sense of excitement among geroscientists and an increasing flood of reports about this in the media.

Geroscience is a young science, and research into ways to slow aging is in its infancy. However, we already know a few things about life extension. We know that the basic processes of aging occur at the cellular and molecular level and that slowing aging requires intervening in those processes, proba- bly using drugs, possibly using genetic engineering (such as the new CRISPR method), and perhaps using stem cell transplants. We know from other medical contexts that those methods tend to be expensive, but we also know that new technologies tend to become automated and less expensive over time. Finally, we know that it's hard to predict new technologies more than 10 or 20 years ahead, for we know that biotechnology has produced things that would have seemed impossible a generation or two earlier. The impossible is a moving horizon.

Still, methods for slowing human aging are years away, and halting aging is well beyond what scientists now forecast. Why write a book about life extension now? Aside from intellectual interest (a good reason by itself), one reason is that we are effectively making policy decisions about life extension right now. We are doing this by making decisions about how much to fund the relevant research. The more aggressively we fund such research, the faster we may learn how to slow aging. Suppose, for exam- ple, that we can discover a way to slow aging enough to gain an extra two decades of life. If funding life extension research aggressively would bring this about 10 years sooner than funding it at current levels, then funding it aggressively adds 20 years of life for those who are just young enough to use life extension technology by then and just old enough that 10 years later will be too late. We don't realize it, and it may seem silly to say this, but we are making decisions about geroscience research funding that may have life-or-death consequences for millions of people. Even if not funding such research beyond present levels is the right research funding decision, we need to think about the ethical issues in order to know that it's right.

Throughout this book I refer to two research funding policies: *Promotion*, which consists of funding life extension research generously enough for it to progress as fast as it can, and *Inhibition*, which consists of funding it at some lower level. (I left out prohibiting such research because I don't think that's feasible and because the arguments for inhibiting it and for prohibit- ing it are much the same.) When I discuss considerations for and against life extension, I tend to present a choice between Promotion and Inhibition. That choice is meant as a stand-in for more general positions for and against life extension.

Very well, life extension is plausible and it's time for a book about it. But why do we need a book about the *ethics* of life extension? What could possibly be wrong with living longer? I will say a lot more about this in the last section of chapter 1, where I survey the ethical issues discussed in this book, but here's a general answer.

Despite its obvious appeal, life extension is controversial. Some bioethicists have serious concerns about it, and surveys reveal that many ordinary citizens share these concerns. (Their concerns are summarized in section 1.9.) Many of these concerns are about whether extended life is a good life—whether you would really like living such a life. Some people believe that extended life would be boring, that living under the deadline of a normal life span spurs us to achievement and makes us use our time well, and that life would lose meaning if it were not bounded by death. Frankly, I think most of these arguments are weak, and they have occupied too much time and attention already. Still, it's natural to begin by asking what it would be like to live an extended life, so I devote chapters 2 and 3 to these and related issues.

There are other issues I find more worrisome. The literature on life extension ethics is laced with comments about unequal access to life extension, dividing society into castes of mortals and near-immortals, making inequality worse, allowing dictators to live forever, and similar concerns. Many people also worry that halting aging, or even slowing it dramatically, will eventually push the world's population into a Malthusian crisis. These topics are where the real problems lie, and I take these concerns very seriously. One of my major aims in this book is to shift life extension ethics away from questions about the desirability of extended life and focus more attention on issues of justice. This is the project of chapters 8, 9, 10, and 11; chapters 2 through 7 set up this project by discussing how life extension affects both those who can get it and those who cannot, and its social consequences for everyone. Chapter 12 covers enhancement issues.

At the end of each chapter from chapter 2 onward, I list the conclusions I believe I've established in that chapter and identify the sections where I argued for them. In chapter 13, I list all the conclusions once again (think of this as an outline of the entire book) and restate the policies I propose concerning (a) whether and how much to fund life extension, (b) how to respond to the fact that many people will not be able to afford access to life extension, and (c) what to do about the possibility that making life extension available may eventually bring about a Malthusian crisis.

In the end I come down in favor of life extension, but only if policies are instituted to prevent a Malthusian crisis. However, I side neither with those who think there are no valid moral objections to it nor with those who think the moral objections are so strong that we should never develop it. I argue that developing life extension is, *on balance*, a good thing and that we should fund life extension research aggressively. However, I also conclude that some of the moral objections—particularly those concerning justice—have real weight. They deserve more careful answers than defenders of life extension usually provide. I hope I've moved that discussion forward.

I thank John Martin Fischer, Benjamin Mitchell-Yellin, Heinrik Hellwig, and the Immortality Project, together with the John Templeton Foundation, for a grant that funded a semester away from teaching to work on this manuscript, and California State University, Fullerton, for funding another semester off for the same purpose. I also thank Michael Cholbi, Steven Munzer, Chris Nattichia, and James Stacy Taylor for helpful comments on earlier versions of some of the chapters. I thank Peter Brinson for his careful reading of the entire text. I owe a debt to Steven Austad and George M. Martin, noted scientists working on aging who happened to be at the University of Washington during my graduate studies in the late 1990s. They probably don't remember me, but I remember them for encouraging my interest and helping me to understand the science of aging. I also benefitted from a recent and very enjoyable conversation about these topics with Aubrey de Grey, who has done more to promote and encourage life extension research than anyone else. I thank Benjamin Mitchell-Yellin (again) and his students for useful feedback on some of the justice issues and for Ben's steadfast encouragement and support. I received helpful feedback on various issues from audiences who heard my life extension ethics presentations at Cambridge University, Halle University, the Brody School of Medicine at East Carolina University, and United World College–USA.

1 An Overview

1.1 Introduction

If you ever wished you could relive your youth with the wisdom you gained in middle age, or wanted more time to accomplish your life goals, or to correct bad choices made long ago, then imagine that medical science gives you that opportunity. Imagine living long enough to gain the wisdom of age while you still have a young body. Imagine having enough time to pursue multiple professions, to see any part of the world that interests you, to *finally* learn a foreign language, to save for retirement over so many decades that you barely notice the expense. Imagine watching the future unfold, not for 70 or 80 years, but for decades more—or centuries beyond. Imagine that.

Life extension has not arrived, but optimism about it has a long history. There's an Egyptian manuscript from around 1600 BCE titled "Book for Transforming an Old Man into a Youth of Twenty." Alchemists and Taoists sought the philosopher's stone to prolong life, and Ponce de Leon sought the fountain of youth. Roger Bacon,[1] Descartes, Benjamin Franklin, and Condorcet all thought it was possible to prevent aging, long before science began to show us how.[2]

That optimism is now grounded in science. The relevant sciences are collectively referred to as *geroscience*, the interdisciplinary scientific study of aging.[3] Geroscience has made breakthrough discoveries within the last 20 years, and many reputable scientists are now publicly saying that life extension is a real possibility. According to some of them, it might be developed within the lifetime of people who are young today.

By *life extension*, I mean slowing, halting, or reversing some or all aspects of aging. Let us be clear: "life extension" does not refer to keeping elderly people alive longer, as we do now when we use respirators and intensive care on aged people in the hospital. Those people age at a normal rate—they just survive longer. Up until now, gains in life expectancy consisted of finding ways to survive longer by eliminating diseases and other things that kill us, not by slowing the rate of aging itself. We are nearing the end of those gains. It has been estimated that eliminating all cancer, heart disease, stroke, and diabetes would increase our average life span from 82 years to 96 years—an increase of only 17 percent.[4] Life extension, by contrast, would alter the *rate* of aging and gain us far more time. If you had life extension that slowed aging but did not halt it, you would live longer and age more slowly than people do now. If you had life extension that halted aging altogether, you would look and feel young indefinitely, at least until you died from something unrelated to aging, such as violence, an accident, or infectious disease. I call this *extended life*, not immortality. If you are *im*mortal, you are not-mortal—incapable of dying. An immortal cannot cease to exist. In extended life, however, your body is biologically normal, apart from aging very slowly or not at all. You can still be mortally wounded.

Many people distinguish between *moderate life extension*, which extends our life expectancy by perhaps a few decades, and *radical life extension*, which extends it by centuries or more. Halting or reversing aging would be the most radical of all. This book covers both moderate and radical life extension scenarios, though I'll tend to talk about radical scenarios more often, partly because they raise the same issues as moderate scenarios but in a starker and more dramatic form, and partly because they raise some issues that moderate scenarios do not.

You might think that developing life extension is an overwhelmingly popular idea, but you would be wrong. Many people favor it, but as surveys of the general public show, there are at least as many people who are not convinced it would be a good thing and quite a few who oppose it. This is not only true of the general public; as we'll see, bioethics thinkers are more or less evenly divided on whether a world with life extension is better or worse than the world we have now.

This book is about the ethics of life extension. Most of the arguments discussed in the life extension ethics literature are arguments *against* life extension (perhaps because postponing aging seems so obviously desirable that it needs no argument), and that is true of most of the arguments discussed in this book. I will argue that most of them are bad arguments—or at least weak. There are some real problems with life extension, but they don't appear until chapter 5, so the first half of this book may give the impression that I think there are no serious drawbacks to life extension. That is not so. I am not among those who think that there are no serious objections to life extension or that those who oppose it are simply confused or uninformed. As you'll see in chapters 6 and 8, I believe that the possibility that life extension will lead to a Malthusian crisis of overpopulation and the injustice of unequal access to life extension are very, very serious problems, and my defense of life extension is qualified by provisos that we adequately address these issues. I also propose policies for doing so. Like many things, life extension is neither a completely good thing nor a completely bad one—there are sound arguments for and against it. As with many things, we must weigh the arguments for and against it and decide where the balance of arguments lies. I will argue that the reasons in favor of life extension outweigh the reasons against it, provided we properly deal with the Malthusian threat and do *what we can* to ensure equal access.

Before we get to the ethical issues, however, we must understand the science behind life extension. That will help us understand why this topic is timely and what life extension might be like. Sections 1.2 through 1.8 are about life extension science. (Those who want to explore that science in more depth will find more material in appendix A; those who don't may choose to skim sections 1.2 through 1.5.) We'll survey the ethical issues in sections 1.9 through 1.11. Thus, this chapter is primarily nonphilosophical in nature. The philosophy begins in chapter 2.

1.2 Optimism about life extension

Well-known mainstream scientists at leading universities are putting their reputations on the line by publicly declaring that we may soon learn how to slow aging.[5] Francis Collins, current director of the National Institutes of Health and former leader of the Human Genome Project, has predicted that the genes involved in aging will be fully cataloged by 2030, with life-extending drugs in clinical trials by then.[6] S. Jay Olshansky, Leonard Hayflick, Bruce Carnes, and 51 other scientists have published a position paper on life extension research in *Science*, *Scientific American*, the *AARP Bulletin*, and the *Journal of Gerontology: Biological Sciences*, saying, "Most biogerontologists believe that our rapidly expanding scientific knowledge holds the promise that means may eventually be discovered to slow the rate of aging. If successful, these interventions are likely to postpone age-related diseases and disorders and extend the period of healthy life."[7]

George M. Martin and the other editors of a special issue of *Science* suggest that "it is through deciphering the biological underpinnings of processes of aging that scientists will likely discover ways to extend the human life-span."[8] Richard Miller, director of the Glenn Center for the Biology of Aging at the University of Michigan, says, "The argument that one can slow aging, and diseases of aging along with it, used to be fantasy, but now we see it like a scientific strategy."[9] The National Institutes of Health (the source of nearly all US government research funding for biomedical science) has formed the NIH Geroscience Interest Group with an eye to treating or curing age-related diseases by attacking the underlying processes of aging itself.[10]

This optimism is not purely speculative; scientists have already slowed aging in several animal species. A mouse in a zoology lab at Southern Illinois University in Carbondale was deliberately mutated to knock out a gene for a growth hormone receptor in order to make it live longer. It died after 1,819 days; in human terms, this is equivalent to living more than 180 years. Single-gene mutations in the *C. elegans* worm produced worms that lived nearly 10 times as long as they do in the wild. Similar results have been achieved with fruit flies. According to Richard Miller, "We now have at least five genes, two diets, and five drugs that extend mouse life spans."[11] As we'll see, the genes manipulated in these experiments are shared by many species, including humans.

Investors have taken notice. In 2000, Cynthia Kenyon and Leonard Guarente founded a biotech company called Elixir Pharmaceuticals to develop drugs to slow our rate of aging by exploiting sirtuin genes. (Guarente suggests they may develop a pill that adds 10 to 30 years to one's life.) David Sinclair also founded a biotech company called Sirtris Pharmaceuticals to develop life extension drugs based on his clinical research with resveratrol and sirtuin genes. In 2008 Glaxo SmithKline bought the company for $720 million.

1.3 Possible methods of life extension and the basic processes of aging

What form might life extension take? Let me start by answering that question on a very general level. For the most part, when we talk about a possible method of life extension, we're talking about two things: the technology we use and the basic process of aging we use it on. In general, we already know that the technologies we might use will include at least one or more of these: drugs, genetic engineering, and stem cell therapy. That said, here are several possible methods, in ascending order of how immediately promising they are (saving the best for last, so to speak). I will indicate which basic process of aging each technology works on. As I explain in appendix A (sections A.3 and A.4), organisms seem to have defenses against the cellular- and molecular-level damage that lies behind what we call aging. This is important, for many of the possible methods might make use of the body's own defenses against aging by improving and intensifying them.

Exotic methods

Some futurists suggest we might achieve life extension with fleets of nanomachines cruising our bloodstream to make repairs, while others suggest we might upload our minds into computers and live forever in cyberspace (or as long as someone keeps the computer running). Perhaps. The future is long, and who knows what might be possible in a century or two? However, I want to stay focused on methods that are closer at hand.

New body parts

Another possibility that gets mentioned now and then is to grow new organs or body parts (using the patient's own cells), then transplant the new part into the patient's body. This is already being done, though the procedure is still in its infancy. However, this is not life extension. Yes, the new organ is brand new, so in a sense that part of you gets 'younger,' but the rest of you is just as aged as you were before transplant. Moreover, it is not clear how much of a patient can be replaced this way; imagine trying to grow and transplant a new brain while preserving your personality, memories, and everything else that makes you *you*. I won't say that making new body parts will not be a *part* of life extension, but it can never be all of it, and it may not be necessary at all.

Telomerase

Still another possible method of life extension involves telomeres—bits of DNA at the ends of our chromosomes. They prevent the ends of your DNA from unraveling, like the caps on your shoelaces. Telomeres get shorter each time a cell divides, and eventually they get *too* short. Telomere short-ening correlates with such age-related diseases as diabetes and cardiovascu-lar disease, interferes with cell division and replacement, and leads to cell senescence and apoptosis (cell suicide).[12]

As it happens, our bodies produce *telomerase*, an enzyme that lengthens our telomeres, but our bodies don't produce enough of it to keep telomeres from eventually getting too short. Some scientists have proposed giving people extra telomerase, thereby preventing telomere shortening. However, there is a big problem with this approach: too much telomerase increases the risk of cancer by inducing cells to divide more often than they should.[13] Perhaps further research will overcome that problem or find some other way to prevent telomere shortening.

Caloric restriction
Caloric restriction is the first successful method of life extension ever dis-covered. The caloric restriction phenomenon was first observed in mice in the 1930s and has been replicated with rats, fish, worms, yeast, and spiders (though the effect does not appear in all species nor in all strains of mice).[14] A caloric restriction diet involves dramatically cutting the number of calories consumed while keeping the diet as nutritious as possible. Ani-mals that are fed a diet that is unusually low in calories but very high in nutrition can live much, much longer than they normally do. Low calories alone will not do the trick, or famine would lengthen lives—the trick is to keep nutrition very high at the same time.[15] Some species live up to twice their natural life span on a caloric restriction diet. Mice, for example, live 40 percent longer on a diet that is 40 percent lower in calories but very high in nutrition.[16]

Many scientists believe caloric restriction could produce similar results for humans.[17] Luigi Fontana of the Washington University School of Medi-cine has put this hypothesis to the test in a study of 22 people who are vol-untarily following a caloric restriction diet for 3 to 15 years and comparing them to 20 control subjects on a normal diet. Those in the experimental group had the cardiac health characteristic of people 20 years younger.[18]

However, caloric restriction does not produce truly dramatic gains in life expectancy. If results with other species are any guide, the best we might hope for is a 40 percent increase—well worth having, but not the radical extension some people hope for. Moreover, some species respond to caloric restriction with a *lower* increase in life expectancy; for all we know, humans might be among them. In addition, the caloric restriction diet is tedious and very difficult to maintain. You have to follow very carefully tweaked recipes and record the nutrients you take in each day so that you can maximize nutrition while minimizing calories. And it makes you relentlessly hungry. (I tried it some years ago but lasted only two weeks. I could almost handle the hunger, but the cooking was an intolerable hassle.)

We might get around these difficulties by developing caloric restriction *mimetics*: drugs that do what the caloric restriction diet does but with a richer diet. Some scientists hope to develop such a drug in the future.[19] That's worth doing, but again, it won't produce dramatic gains in life expectancy.

Stem cells

A more radical possible method involves stem cells. A stem cell is a kind of cell that can produce new cells and, ultimately, new tissues for the body. Because they can produce many different kinds of cells, they are called *undifferentiated* cells. Every embryo starts out as a collection of nothing but stem cells. As the embryo develops, the cells begin to differentiate into different types of tissues. We also have stem cells as adults. Adult stem cells don't produce as many different kinds of cells as the stem cells found in embryos, but they can produce *some* kinds of cells, and they replenish and renew tissues in our body. In that way they defend against aging, or at least some aspects of it.

Adult stem cells maintain tissues up to and during an organism's peak reproductive years. However, after the peak reproductive years, adult stem cells cease doing this, and cell and tissue health go into decline. The regenerative capacity of adult stem cells is compromised over time by genetic mutations, epigenetic changes (things that switch genes on or off), and changes in the environment constituted by the body itself. As a result, older people have fewer adult stem cells than younger people, and stem cells work less well as we age. We don't yet know whether stem cells repair all types of age-related damage or just some of it, but it's possible that aging results at least in part from having fewer and less effective stem cells over time.[20]

If we could increase the number and potency of our stem cells, we might slow some aspects of aging and even repair some age-related damage. The technology is already partly here, for we use stem cells in some cancer treatments. Cancer patients who undergo chemotherapy and radiation therapy often lose stem cells in their bone marrow; one therapy for this involves taking new stem cells from the blood or bone marrow of a donor and injecting them into the patient. To make this a practical life extension method, we would have to find ways to produce more stem cells using the patient's own tissue (matching donors are hard to find). Perhaps we can take some from the patient and clone them to produce more. If so, stem cell therapy might turn out to be an important life extension method, though whether life extension can be done with stem cells alone is another question, for there may be some basic processes of aging that stem cells don't help defend against.

Genetic interventions: drugs and genetic engineering

Finally, we might (and probably will) intervene in the genes and genetic pathways that regulate aging. A genetic pathway is a set of genes that interact with each other to perform some function, typically involving cellular structures and chemical interactions. Scientists have so far discovered more than two dozen genes involved in aging that are shared by many species, including mice, rats, fruit flies, nematode worms, yeast, and many other multicellular organisms.[21] Some researchers have slowed aging in various laboratory animals by tweaking these genes.

The genes and genetic pathways that regulate aging may be among what I call the "maintenance systems," which repair aging-related damage and prevent us from aging too fast (see appendix A, section A.4). These are the body's built-in systems for defending against and repairing the cellular- and molecular-level damage that initiates aging. If the body can *already* slow aging and repair its damage to some extent, perhaps we can tweak those systems to make them work even better. Steven Austad has suggested reengineering our genes to produce more antioxidants to combat aspects of aging that are traceable to oxidative stress, and William Clark speculates that gene therapy might enhance the cell's natural means of repairing molecular and cellular damage associated with aging.[22]

There are two ways to manipulate the genes and genetic pathways that regulate aging: drugs (I use that term rather broadly) and genetic engineering.

We can use drugs or proteins to *modulate* the genes that regulate aging (to "modulate" a gene is to regulate or alter its operations), thereby slowing aging. Genes perform their functions when they are *expressed*. A gene is expressed when the information coded in the gene is used to produce a protein or an RNA molecule to carry that information elsewhere in the cell, triggering some physical event in the cell and ultimately producing whatever consequences that gene is responsible for. Some drugs can alter a gene's expression—switching it on or off, so to speak—either causing the gene to produce its protein or RNA molecule or preventing it from doing so. If we can find the right drugs, we might switch the genes that control aging on or off, as needed. We might also find ways to alter gene expression by inserting the proteins themselves into the body.

Some drugs that may modulate genes involved in aging are now undergoing clinical trials for age-related diseases such as cancer or diabetes;[23] they might be used to slow our rate of aging as well. Because there are many different genes and genetic pathways involved, we will probably find partially effective drugs that slow aging to some extent long before we find drugs (or any other technology) that halt or prevent aging altogether. If radical life extension arrives, it will arrive in stages.

Genetic engineering modifies the operation of genes in a more permanent way. Until recently, genetic engineering was extremely difficult to do, but there is a new method of genetic engineering that works with far greater precision, efficiency, and speed than previous methods. This new method exploits a cellular mechanism called CRISPR (Clustered Regularly-Interspaced Short Palindromic Repeats). Bacteria that have this mechanism can make copies of viruses and insert those copies into a cell's DNA so that the cell can later recognize those viruses and defend against them. This is a naturally occurring form of genetic engineering, and scientists have recently learned how to use it do genetic engineering of their own. It's far easier, far more effective, and far more precise than older methods of genetic engineering. Thanks to CRISPR, genetic engineering may be on the verge of becoming an important medical tool.

Whether we use drugs or genetic engineering to modulate gene expression, we must first identify the genetic pathways involved in aging. The three most promising possible genetic pathways for regulating aging include one that involves the TOR gene, one that involves sirtuins and NAD precursors, and one that involves insulin and insulin-like growth factor (IGF). We have some promising drugs for the first two genetic pathways but not yet for the third.

The TOR gene, rapamycin, and metformin

This genetic pathway involves a protein called TOR and the gene that produces it. The TOR gene is sensitive to food supply: when there is a lot of food, TOR activity increases, but when food is scarce, TOR activity declines and the TOR gene directs cells to put metabolic resources into repairing DNA, making mitochondrial components, and performing other kinds of cell repair rather than putting resources into growth and reproduction. (This may lie behind the caloric restriction phenomenon.)[24]

We have a drug for the TOR pathway. It is called rapamycin (*TOR* stands for "target of rapamycin"), and it is made from a bacterium found in the soil of Easter Island (*Easter Island* is also called "Rapa Nui"). Rapamycin suppresses TOR activity, and that in turn stimulates cell repair. In other words, rapamycin appears to ramp up some of the body's natural defenses against aging. Even late in life, taking rapamycin can extend life span, reverse cardiac decline, and improve immune function in mice, and a recent study showed that it significantly boosts immune function in the elderly.[25] Mice given rapamycin have maximum life spans that are 12 percent longer than normal.

Unfortunately, rapamycin can also impair the immune system and impede wound healing. However, there may be a safer alternative: metformin, a drug currently used for diabetes.[26] Like rapamycin, metformin inhibits TOR activity. It also activates an age-related enzyme called AMPK (which is also stimulated by caloric restriction), so metformin may be a caloric restriction mimetic as well.[27] Metformin has been shown to improve health span in mice and may even lengthen their life span.[28]

Sirtuins and NAD precursors

Until recently, genetic pathways involving enzymes called "sirtuins" looked very promising. Resveratrol, found in red wine, among other things, revs up the activity of "sir" genes and was considered the basis for an antiaging drug. Unfortunately, recent studies have convinced many researchers that resveratrol and sirtuins are not involved in slowing aging after all.[29]

However, sirtuins are still in the running, partly because of recent discoveries about the relationship between sirtuins and NAD precursors. In 2000, MIT biologist Lenny Guarente discovered that a molecule called NAD (nicotinamide adenine dinucleotide) increases the activity of sirtuins. Another molecule, called NMN (nicotinamide mononucleotide), boosts NAD levels.

Guarente's protégé, David Sinclair, recently discovered that injecting NMN into elderly mice for a week restored their mitochondria to a youthful state. (Mitochondria are the parts of our cells that provide energy.) NAD precursors like NMN also improve cognitive function in mice, and some other sirtuin activators improve health span and slightly extend life span in mice.[30] NAD concentration declines with age and may contribute to age-related diseases. NAD supplements might help restore mitochondrial function.[31] However, increasing NAD levels might also promote tumor development.[32]

Insulin/IGF

Another promising genetic pathway involves insulin and insulin-like growth factor (IGF). According to some scientists, when insulin and IGF levels are high, organisms put their resources into growth and reproduction, but when they are low, organisms put their resources into maintenance and repair. Insulin and IGF are hormones that regulate metabolism and reproduction, and some say that caloric restriction works by reducing signaling in the insulin and IGF pathway. One of the genes regulated by the insulin/IGF pathway is FOXO, a gene found in many exceptionally long-lived people. Activating FOXO boosts the immune system and increases antioxidant production. We don't yet have a drug for this pathway, but identifying the pathway is a first step toward developing such drugs.

1.4 Is it possible to reverse aging?

Some people want to go further. They want to *reverse* aging and become young again. There are tantalizing recent news announcements to the effect that scientists have effectively reversed some effects of aging in various organisms. A team of Japanese scientists at the University of Tsukuba, working with two genes in the mitochondria of human cells, have discovered how to switch them on, thereby restoring the cell's capacity for respiration to that of younger cells; they interpret this as *reversing* aging in the cell line.[33] Researchers at the University of Heidelberg have restored lost memory to mice by giving them extra DNA methyltransferase.[34] Scientists at the Albert Einstein College of Medicine have reversed a decline in immune function in mice by giving them antioxidants.[35] Scientists at the University of Pittsburgh have genetically engineered some mice to age faster than normal, then injected them with muscle cells from young mice;

the injected mice became much healthier and their stem cells began dividing faster, as if they were younger.[36] According to a recent review in *Science*, "There is some evidence that [reversing aging] may be achieved by some interventions that target mechanisms of aging. For example, mTOR inhibitors such as rapamycin . . . can partially rejuvenate immune stem cell and cardiac function in mice and can perhaps also restore immune function in elderly people too."[37]

Journalists tend to characterize such announcements as instances of "reversing" aging, but reversing aging sounds impossible, as if it requires making time go backward. However, there is a sense in which simply *slowing* aging is a reversal of aging, for it resets your life span, giving you the life expectancy you had at an earlier age—just as if your age had been lowered somehow. As we'll see in appendix A, section A.1, most geroscientists define aging in terms of increasing risk of failure: you are more aged the higher the odds are that your tissues and organ systems will fail. When you are young, the odds are very low; when you are elderly, they are very high. Life extension, then, involves lowering the risk of failure.

Of course, this leaves something out. We would not say aging had been reversed if the risk of failure in a 60-year-old were lowered to that of a 30-year-old but the 60-year-old still looked and felt 60. That might count as reversing aging in a technical sense, but it's not quite what we want. To reverse aging in the full sense requires removing or correcting the manifestations of aging so that the 60-year-old looks and feels like a 30-year-old. Presumably this would require removing or correcting the manifestations of aging down to the cellular and molecular level.

I will hazard a speculation here. Perhaps there is a connection between reducing the risk of failure and making an organism look and feel younger. What is involved in reducing the risk of failure to what it was at an earlier age? It may include repairing all the cellular- and molecular-level damage involved in the basic processes of aging. After all, if we don't do that, how would the risk of failure get any lower? Surely some repairs are required at that level just to slow the risk of failure, for the cellular and molecular-level damage that lies behind aging occurs throughout our lives. If we improve those repair processes, would that also repair some of the outward manifestations of aging that damage has already caused, such as reduced immune system response, dementia, bone tissue loss, skin wrinkling, and so on?

Perhaps; the animal experiments mentioned above suggest something like this might be possible. For now, we don't know whether reversing aging is possible, but we can't rule it out.

1.5 Why slowing aging might be harder than we realize

This book is premised on the claim that life extension is possible and that it's likely we will eventually develop new technologies that slow, halt, or even reverse aging. However, there are reasons to think this might be harder to achieve than some science writers have suggested, and it would be irresponsible to gloss over the challenges.

First, the research process itself faces some obstacles, though these are the easiest ones to overcome. For one thing, there is the time involved. Research on methods of life extension will require identifying biomarkers—biological indicators to show how much and how fast someone is aging—unless we want to research life extension by following individuals through entire life spans, which may take a century or more.[38] (Scientists are already working on this.) For another, research in the United States is inhibited by the fact that the Food and Drug Administration does not currently have a regulatory pathway for interventions meant to slow or halt aging, so it is hard to get FDA approval for clinical trials on human subjects. However, it may be possible to get around this by testing interventions meant to treat age-related diseases (even if they have the side effect of slowing aging) or using animal studies.[39]

A much more serious problem is that there might turn out to be so many genes and genetic pathways involved in aging that devising drugs to slow or halt those processes is too complex for any foreseeable technology. This is more likely if the mutational accumulation theory of aging (see appendix A.4) is correct, for we would then have many unrelated genetic defects to contend with rather than a smaller set of related pathways. Even if the mutational accumulation theory is mistaken, we still have many redundant gene functions and complex connections between genes and gene pathways, and it may be very difficult to map the entire system well enough to safely modify it. In light of these complications, it may be a long time before we develop safe and effective life extension methods.

1.6 Are any life extension methods available right now?

If you want to practice life extension right now, you have a few options, though there is no scientific consensus about whether they work or how well. First, you can adopt a caloric restriction diet. There are books that tell you how, and they include recipes.[40] Keeping nutrition high enough is a challenge and requires carefully tweaked recipes. (Simply cutting calories is not enough—this is not a weight loss regimen, though you *will* lose weight.) There is software you can use to keep track of your diet to make sure your calories are low enough *and* your nutrition is high enough.[41] There's no scientific consensus on whether this diet works for humans or how well it might work, but it shows promise.

You can also buy two different nonprescription commercial supplements that provide, in pill form, some of the potentially life-extending substances geroscience has recently discovered.[42] So far there are no published studies on the safety or efficacy of these products.

One of these products has been on the market for some time as a vitamin supplement for conventional purposes. It's called "Niagen," and it's made by a company called ChromaDex (the company's board of science advisors is chaired by Roger Kornberg, who won the Nobel Prize for chemistry in 2006). Niagen is a form of vitamin B3 (niacin); it's also a form of nicotinamide riboside (NR). NR is a NAD precursor, a molecule that boosts NAD levels. NAD, in turn, boosts sirtuin levels. High sirtuin levels have been shown to have some impressive life extension results in mouse studies (see section 1.3). Some people are now using Niagen in hopes that it will slow at least some aspects of aging. ChromaDex now advertises it for that purpose and announced that it has conducted a clinical trial of Niagen on human subjects that demonstrates that a single dose of NR caused significant increases in NAD levels. That study has not been published yet. Niagen is available for $46.99 a month, plus shipping and handling.

ChromaDex now has a competitor. In early 2016, Lenny Guarente, a biologist at MIT and a leading geroscientist, announced that Elysium Health—a company he cofounded—will be selling an NR supplement as a product it calls Basis. (Elysium Health's board of science advisors includes five Nobel laureates and other leading scientists.) In addition to NR, Basis also includes pterostilbene, a substance similar to resveratrol. Elysium believes that NAD boosters work synergistically with resveratrol to improve mitochondria function and defend against age-related diseases. Moreover, pterostilbene is said to be more powerful than resveratrol. Basis is available online for $50 a month.

Novartis, a drug company based in Switzerland, does not have a life extension product on the market yet, but Novartis is looking for alternatives to rapamycin that it can market as a life extension product.[43] Rapamycin appears to improve the body's natural defenses against aging (it works on the TOR pathway) and has produced striking results in studies on mice. However, it's also very expensive, and it's used as a powerful immune suppressant, which makes it dangerous. We should not be surprised to see Novartis market a life extension pill based on some substitute for rapamycin.

If you don't want to wait for Novartis's alternative to rapamycin, you can get a prescription from your doctor for metformin, a diabetes drug that is now off-patent and available as a less expensive generic drug. Metformin produced impressive results in some life extension studies on mice, and it inhibits TOR activity. Some people are now taking metformin as a life extension supplement.

1.7 How long would we live?

How long would we live? No one really knows. Steven Austad believes that the first person who will live to be 150 is alive today, but has not offered a longer forecast than that.[44] Richard Miller says we might increase our life span by 40 percent, to an average life span of 112, with some living to 140.[45] S. Jay Olshansky believes any estimate past 130 is "ridiculous."[46]

Other scientists endorse more dramatic forecasts. Tom Johnson, who works with nematode worms at the University of Colorado, says that "if humans are as malleable as worms, we could see life-spans of 350."[47] Cynthia Kenyon says there's no theoretical reason why aging in worms can't be halted altogether, though she hasn't extended this claim to humans.[48] The most optimistic forecasts emerged from a group of 60 demographers, gerontologists, and other scientists who research aging, who were asked how long a child born in 2100 can expect to live. The average of their answers was 292, with some estimates ranging as high as 5,000 years.[49]

Assuming we slow aging but do not halt it, we will still have such a thing as old age. However, we don't know what that will look like. One possibility is that senescence—what we think of being as being frail and elderly—will occupy the same percentage of a life span that it does now. In this scenario, if you lived to 170 rather than 85, you would age twice as slowly and the phase we call "old age" would be twice as long as it is now: 30 years instead of 15, perhaps. Another possibility is that life extension might produce *compressed morbidity*. In that scenario, you would be fit and healthy during almost all your life and then age and die in a very short time: if you lived to 170, your "old age" might last the usual 15 years—or even less. Of course, if we halt aging altogether, there will be no senescence at all.

Suppose we could halt aging entirely and never aged at all after, say, age 25—roughly the age at which we start aging. How long would we live then? Not forever. A cure for aging is not immortality. Disease and accidents will still take a toll, so we can answer this question by asking what the average life expectancy at birth would be if people died only from causes that are completely unrelated to aging. On one estimate, the average life span would be 1,200 years.[50] Shahin Davoudpour, a demographer who helped me with chapter six, estimates it would be approximately 1,000 years. That's the figure I use in this book. Bear in mind that this is an average; individuals might live far longer or far shorter lives than that, depending on luck and caution. Of course, these estimates work with the accident rate as it is now, and we might reduce the accident rate so that lucky or nimble people would live far longer than that.

1.8 Misconceptions about what life extension would be like

Now that we have a sense of what life extension might look like, we can clear up some misconceptions about it. First, because aging involves several basic processes and perhaps thousands of genes, life extension probably won't take the form of a single pill, and because aging is not caused by a virus or bacteria, it won't take the form of a vaccination either.

Second, you will not wake up one morning, check the news, and see that life extension has been invented. Life extension will probably require several different interventions working on various genetic mechanisms. These interventions will appear one by one over time. They may work on some aspects of aging and not others. Later versions may be more effective than earlier versions. The various interventions will be developed at different times, and this will take years. There may be unforeseen side effects. In short, we'll see components of life extension coming over the horizon long before it all arrives in a complete form.

Some proponents of life extension speak of *longevity escape velocity*, which happens when advances in life extension arrive often enough that your remaining (not merely total) life expectancy increases with time. In other words, if you live long enough to get the earliest life extension remedy and add an extra decade or two to your life, then you'll live long enough to get the next, improved version of life extension, adding another few decades, and thereby live long enough to get the kind of life extension that stops aging completely. Maybe. That's a lot of trains to catch, and one of them might leave its station after you've run out of time to wait. Moreover, you've got to have enough money for the tickets. Still, there's room for hope.

Third, there's little reason to think that using life extension will be an irreversible decision. This is important, for many depictions of life extension present it as something that cannot be reversed or undone. Think back to the scenarios we all know from mythology and literature where someone receives eternal life at one stroke from a god, a genie, or a potion, only to find that he or she cannot die and that immortality becomes boring, meaningless, and, finally, a living hell. (Simone de Beauvoir's *All Men Are Mortal* is a good example of this genre.[51]) Such stories have a moral: Be thankful you age and die. If you were wise, you wouldn't want it any other way. (Giving that warning is a way to seem wise.) The lesson is that we should think long and hard before commencing an endless life we might not want to be stuck with.

These stories color our sense of what life extension will be like, and they color it wrong. Using life extension will almost certainly be a reversible choice. Unlike vaccinations, drugs have to be taken on an ongoing basis. This is probably true of stem cell transplants too, for transplanted stem cells will eventually age and require replacement in turn. Genetic engineering is permanent in the sense that it does not have to be repeated, but in principle, this should be reversible too. (If we can splice DNA in, we should be able to remove it. If we can remove DNA, we should be able to put it back in.) Moreover, if you need drugs or stem cell transplants in addition to genetic engineering, as seems likely, then going off those drugs or ceasing the stem cell treatments should be enough to resume aging even if your genetic engineering stays intact—otherwise you wouldn't need those things in addition to genetic engineering in the first place. Finally, extended life is not immortality; you can still die from causes unrelated to aging (including self-inflicted ones).

In short, you should be able to reverse the decision at any time by going off your life extension meds, so to speak, and resume aging until you die. This means you can try extended life for a while, see how you like it, and leave it if you don't.

And now we're ready to see why some people think it might be better not to give the human race that choice.

1.9 Not everyone thinks life extension is desirable

The reasons for extending your life are obvious. If your life is good, it's good to have more of it. You would have more time to accomplish your life goals and contribute to the world. You might live long enough to see new technologies that enhance your cognitive and physical abilities and enable you to transcend current human limitations. A longer life span might give you greater wisdom and maturity, and if enough of us undergo this, the human race might finally grow up, reducing pettiness, superficiality, and greed. A world of such people might be less violent, less prone to war. They might value the future more and take a long-range view when it comes to the environment and the welfare of future generations.

If nothing else, life extension is the best way to prevent age-related diseases such as heart disease, cancer, stroke, Alzheimer's disease, and diabetes. According to the Centers for Disease Control, more than 57 percent of the 2,596,993 deaths in the United States in 2013 were caused by one of these diseases:[52]

Heart disease: 611,105
Cancer: 584,881
Stroke: 128,978
Alzheimer's disease: 84,767
Diabetes: 75,578
Total: 1,485,309

Slowing or halting aging will not prevent *all* cases of these diseases; after all, young people get cancer too. It will, however, reduce the incidence of these diseases, and it would eliminate any diseases whose cause and development involve basic aging processes themselves, such as type 2 diabetes and osteoporosis. Some advocates of life extension speak of a "longevity dividend" here. The *longevity dividend* includes preventing age-related diseases and preventing the ailments and frailties of age. It also includes increasing the number of years that people can work, easing the pension crisis, and so on.

Even so, many people don't want life extension. Some don't want it for themselves, and some don't want anyone to have it. In 2013, the Pew Research Center's Religion and Public Life project contacted 2,012 American adults to learn their views concerning human life extension. Seven percent of them had heard or read a lot about it and 30 percent had heard a little, but more than half had never heard of it before taking the survey. Of this group, 56 percent said they would not want life extension if it were available to them, but 68 percent of them thought most people would want it. (Let me add a personal observation: given that more than half of this group had never heard of life extension, I question how well they understood it. In my experience, many people think life extension involve keeping elderly people alive longer, as we do in hospitals now. If it were like that, I wouldn't want it either.) Fifty-one percent thought life extension would be a bad thing for society; 41 percent thought it would be good for society.[53] (On the other hand, in a 2012 study of 1,231 Canadians, 59 percent said they supported moderate life extension of up to 120 years.[54])

These results are very similar to the findings of a survey conducted by two universities in Australia, where 605 adults were asked to think about slowing aging enough that people could live longer than 150 years. Of this group, 65.1 percent said they favored research into life extension technologies, but only 35.4 percent said they would take life extension pills if they could. Many of them (38.9 percent) said they thought extended life would do them more harm than good.[55]

Why would anyone reject extended, youthful life over the death sentence we call a normal life span? Some of this may be anxiety in the face of radical change. I respect that anxiety; if it comes, life extension will be one of the biggest changes the human race has ever seen, comparable to the domestication of animals, the development of writing, and the rise of agriculture, cities, science, and industry.

Some of this may be an *adaptive preference*—sour grapes. In Aesop's fable, the fox that couldn't reach the grapes decided the grapes were sour and that he didn't want them anyway. He formed an adaptive preference. Adaptive preference formation happens when you adapt what you *want* to what you can *get*. Because you can't reach the grapes, you decide you don't want them anyway (but you would want them if you believed you could get them).[56] Aging and death have always been unavoidable, and perhaps we adapt to that fact by convincing ourselves that this is a good thing.

Some of this may be covert envy. Life extension may not arrive for another generation, and many of us are neither young enough nor rich enough to get it. Most of us, including me, will age and die while watching life extension come closer and closer to reality or, worse, live just long enough to watch a lucky few get it. It is natural to envy those who get it, and some of us may sublimate our resentment under a posture of opposition.

None of these psychological factors is an argument against life extension, but they may incline us to find arguments against life extension more plausible than they really are. (Aubrey de Grey calls this, with some justification, the "pro-aging trance.") That said, some of those arguments have some weight, and members of the public with no formal training in philosophy or ethics have done a good job of identifying many of the main arguments on both sides. To see those arguments, we turn to another study conducted by the Australian researchers, who asked 65 people in focus groups to identify the most important ethical issues raised by life extension.[57]

Some of them argued that aging is a disease to be cured. Some argued that inhibiting life extension research is morally equivalent to killing all the people whose lives would be longer if life extension were made available sooner. Some argued that a society of people with longer lives would be more knowledgeable and more responsible about long-term interests and the long-term consequences of their actions.[58]

But many others thought life extension might not be a good thing. Some worried about "the potential for boredom at advanced ages . . . and the prospect of outliving loved ones." Some worried that slowing human aging was contrary to God's plan and "unnatural" in a morally bad way. Others worried that we would miss out on the benefits of growing old. Some said that aging is not a disease, and therefore we should not cure it even if we can.

Many of them said that money should be spent on other social needs (especially the conventional medical problems of the poor) before spending anything on life extension. Many worried that widely available life extension would cause a Malthusian crisis of population pressures, resource shortages, and pollution. (This concern was also acknowledged by some people who favor extending human life). Many also worried that life extension would be available only to the rich and that it would make existing inequalities even worse as the rich consolidate their position by living longer. Others replied that new technologies become cheaper over time. There were concerns about discrimination against younger people, opportunities for advancement drying up as older people hung on to their jobs and positions, and a general ossification of thinking as innovation and social change slowed down.[59]

The ethical concerns expressed by American adults in the Pew Research survey resemble those expressed in the Australian surveys. Fifty-three percent of the American adults surveyed thought the economy would be more productive if life extension were widely available. However, 66 percent of them thought only the wealthy would have access. Another 66 percent thought widespread use of life extension would strain the environment, presumably through increased population pressures. Fifty-eight percent thought life extension was "fundamentally unnatural." (You might think these views correlate with religion, but the survey indicates otherwise; religious affiliation made little difference to one's views here.)[60]

There is a similar split of opinion among academic specialists in ethics. The *prolongevists*, as they are sometimes called, favor developing life extension and making it widely available, partly for the same reasons expressed in the surveys. They include John Harris, Christine Overall, Julian Savulescu, Arthur Caplan, and me, among others. The *antilongevists* include Leon Kass, Daniel Callahan, Bill McKibben, and Francis Fukuyama, among others; their concerns closely track those of adults in the surveys who oppose life extension. Of course, the prolongevist/antilongevist distinction is too simple, for most prolongevists acknowledge some potential problems with life extension or qualify their support by calling for certain safeguards against those problems (as I do). I also suspect that most antilongevists would concede the advantages of extended life, though they seldom say so in their writings; they simply believe that the drawbacks to life extension outweigh the advantages.

1.10 Why worry about this now?

Mainstream scientists tell us that some form of life extension may be developed within this century. However, most of that century lies ahead, and most policy decisions about life extension lie there too. Why worry about this now? Why not think about it later on, after life extension has been developed and we understand it better?

We need to think about this now because we *are* making an important policy choice right now: how much to spend on life extension research. Here are the main alternatives:

Inhibition
We can try to inhibit such research—to slow it down. We might refuse to spend public funds on such research or fund it only very conservatively, or try to prohibit it temporarily, until we develop a complete regulatory framework or a strong social consensus about such research.

Prohibition
We can try to prohibit anyone from researching and developing life extension. (This is not the same as prohibiting anyone from using it once it has been developed; that is an access policy, not a research policy.) To the extent that this is not feasible, Prohibition amounts to a kind of Inhibition in practice.

Promotion
We can fund life extension research aggressively in order to develop life extension methods as quickly as possible.

Choosing among these policies has huge consequences. If life extension is possible, then any policy under which life extension becomes available later rather than sooner is lethal for those born too soon. If we delay the advent of life extension by one generation, then there is one future generation whose members (at least the prosperous ones) will die years or decades earlier than they might have, having died just a generation too soon to get the cure. That is a staggering loss of life—or at least life-years.

Our current life extension research policy is Inhibition by default, for life extension research is funded at a low level compared to research funding for most medical conditions. To see this, consider what Promotion might cost. A group of scientists led by S. Jay Olshansky have called for a "Manhattan Project" of antiaging research called the Longevity Dividend Initiative.[61] They call on Congress "to invest $3 billion annually to this effort, or about 1% of the current Medicare budget of $309 billion."[62] I am not sure how they arrived at a figure of $3 billion annually, but perhaps they noticed that this is roughly what the US government spends researching a disease when it takes the disease seriously. Most biological and medical research in the United States is funded by the National Institutes of Health (NIH), an arm of the US Department of Health and Human Services. In 2014, the NIH spent $7.324 billion on genetics research, $5.392 billion on cancer research, $5.002 on infectious disease research, $3.639 billion on "rare disease" research, and $2.978 billion on HIV/AIDS research, to take just a few examples.[63] Notice that those sums range from about $3 billion a year on up.

Here is another way to look at this: the major diseases of old age are cancer, heart disease, stroke, and diabetes. The NIH estimated that in 2015, it would spend $8.763 billion on those four conditions.[64] Because those diseases are associated with aging, slowing aging may be an effective way of reducing their incidence. We are spending just under $9 billion researching these diseases but (as I show below) only a tiny fraction of that sum researching the aging process they correlate with. So although I have not seen the budget behind the $3 billion figure, it seems likely that researching life extension properly is at least as expensive as researching a single disease like HIV/AIDS.

So how much is the NIH spending on life extension research? In 2012, the NIH spent $2.493 billion on aging research, but most of that was for age-related diseases and conditions of the elderly rather than research on the basic biology of aging—the kind of research that might have implications for life extension. For 2017, the National Institute of Aging (NIA) requested $183.174 million for research on the "basic biological mechanisms underlying the process of aging and age-related diseases" in order to "provide the biological basis for interventions in the process of aging, which is the major risk factor for many chronic diseases"[65] Intervening in the process of aging sounds like life extension. In addition, the NIA requested $132.932 million for "Intramural Research" on a variety of issues; some of those issues are related to life extension in one way or another. Thus, the NIH appears to be spending somewhere between $183 million and $315 million on what might be considered life extension research. That's a lot, but even $315 million is less than a tenth of what the NIH spends every year on "rare diseases" alone. By no stretch can we call that a funding policy of Promotion.

What about the pharmaceutical industry? Drug companies are not interested in investing in life-extending drugs unless the drugs are considered prescription drugs by the US Food and Drug Administration. The FDA will not grant prescription status unless the drug treats a disease, and the FDA does not consider aging a disease.

What about the rest of the private sector? Several wealthy individuals and foundations, particularly in Silicon Valley, are trying to take up the slack. The best-known foundation is the SENS Research Foundation, headed by Aubrey de Grey. However, so far as I can tell from their websites, the total reported spending by all these groups appears to be no more than $100 million over the last decade and a half (for that entire period, not per year).[66] (I would be delighted to learn that it's much larger, but that's all I can glean from their websites.) The National Academy of Medicine, a private nonprofit organization, will award at least $25 million for research aimed at slowing aging.[67] Unity Biotechnology has raised $116 million from investors including Jeff Bezos of Amazon and Peter Thiel of PayPal.[68]

There is one outlier whose funding may dwarf all others: in 2013, Google founded a company called Calico (California Life Company) to research ways to slow aging and prevent age-related diseases—with an initial investment of $1 billion.[69] Google has not announced how much money it will put into Calico beyond the first billion, though it did announce on September 3, 2014, that it would partner with AbbVie, a pharmaceutical company, to fund a research and development center with initial investments of $250 million each and potential further contributions of $500 million each.[70] That is a potential additional funding of $1.5 billion (Calico has announced several additional partnerships with other institutions without disclosing the funding). However, that $1.5 billion is not a per-year figure and thus nowhere close to $3 billion per year for life extension research.

In short, we have a research funding policy of Inhibition. Even if Inhibition *is* the right policy, it is wrong to institute Inhibition simply because we have not thought about it. There is too much at stake to be that thoughtless. We must choose our research funding policy consciously. To do that, we must think about all the ethical issues life extension raises.

1.11 A survey of the moral issues

My discussion of life extension ethics is framed around three groups. The *Haves* are those who have access to life extension (they can afford it) and they get it. They live extended lives. Then there are the *Will-nots*; they too have access to life extension (they are just as well-off as the Haves), but unlike the Haves, they turn it down. Finally, there are the *Have-nots*—all those who do not have access to life extension.

Chapters 2 and 3 are about the Haves—specifically, about what it would be like to live an extended life and whether extended life is good for you. Most of the issues in these chapters arise—to varying degrees—both in moderate scenarios where aging is slowed enough to extend life by a few decades and in radical scenarios where aging is completely halted and life is extended by centuries or more. Obviously, these issues weigh more heavily the longer you live. In chapter 2, I discuss Bernard Williams's argument that in a sufficiently long extended life, you'll suffer unbearable boredom unless you change enough along the way—and if you change enough, you change so much that you eventually become a different person (and thereby cease to exist, in a sense). Concerns that extended life will be boring or pose a threat to our identity over time are quite common. Williams is my vehicle for addressing this. I consider various objections to Williams, endorse some of them, and add a couple of my own, concluding that these concerns are not good reasons to refuse extended life.

In chapter 3, I discuss other concerns about extended life. Bioconservatives have argued that being mortal is good in many ways and that extended life is bad for us. At the risk of sounding flippant, I call those advantages the "death benefits." The argument is that extended life might make it harder to accept death, easier to procrastinate and waste time, and harder to develop the virtues that arise when we grapple with mortal threats. If you extend your life, you lose the death benefits—or so some bioconservatives argue. I argue that some of the reasons behind this worry are valid but very weak and the rest are not good reasons at all, even in a life that spans centuries.

In chapter 3, I also discuss how life extension changes the human condition for those who live extended lives. Here I focus entirely on radical scenarios where aging is completely halted, though these issues also arise in a diluted form in moderate scenarios. In the new human condition,

(a) death will be unscheduled, (b) your time of death and your life span will be indefinite and unpredictable, and (c) when you die, you will lose far more time than death takes from us in a normal life span. Finally, (d) all of this is almost certainly elective; in the most plausible scenarios, you can always go off your life extension meds and resume aging if you tire of extended life. This new human condition poses new challenges in coping with death.

The Will-nots are an overlooked group in the life extension literature, and in chapter 4, I discuss some issues unique to them. The Will-nots can afford life extension but turn it down. First, we need to ask whether turning it down, or terminating it after you start it, is a kind of suicide. In other words, does the fact that life extension is available force a Will-not to choose between an extended life she doesn't want and what she may regard as a kind of suicide? My answer, despite my pro-extension stance, is "yes." This is suicide, or at least the moral equivalent of suicide, but it is not, I argue, an *immoral* kind of suicide. Some suicides are morally permissible, and this is one of them. That, however, is not the end of this issue, for there is a kind of "moral injury" in being forced to do something you sincerely believe is immoral, even if your moral belief is false. That too counts as a kind of harm.

Second, we need to ask whether the supposed benefits of normal aging (the death benefits) are reduced by the fact that the Will-nots *could* extend their lives and that normal aging is, for them, elective. For example, mortality may well spur us to use our time well when we have no alternative, but does it do that just as effectively when we know we can have extended lives if we wish?

Chapter 5 is about all three groups; it concerns the effects of widespread life extension on society as a whole. Some consequences will be good. For example, we may find that a larger portion of the human race takes the long view when it comes to protecting the environment and that the human race becomes wiser, more mature, less superficial, less violent, more informed, better educated, and more responsible. The human race might finally grow up. And there is that longevity divided.

Some consequences will be bad. We may face what I call *entrenched elders* and *opportunity drought* (how can young academics get jobs when old professors hang around forever?). Perhaps an extended world would suffer a slower rate of innovation and social change. Entrenched political power is particularly worrisome: What are the odds of a revolution in North Korea if the Dear Leader never dies?

In chapter 6, I discuss the most important potential bad consequence: the threat of a Malthusian crisis. If many people live for centuries, over-population and demands on the environment and natural resources may become much worse. It's not clear how likely this is, but we should take this threat seriously. I propose a policy I call Forced Choice, which limits the number of children for those who extend their lives. There are many practical problems for such a policy; I suggest some ways to solve those problems and try to show that such a policy can be feasible and just. This chapter contains several graphs and projections I developed with Shahin Davoudpour, a demographer, which demonstrate what level of fertility (that is, how many children per couple) we need to aim at in order to have life extension while avoiding a Malthusian disaster.

The *Have-nots* present very different issues, and chapter 7 concerns two harms that are specific to them. The first harm is fairly obvious: they will suffer "distress"—a generic term for all the negative emotions and undesir-able feelings they experience when they compare themselves to those who get life extension. But do they really have something to be distressed about, just because someone else gets to live longer? We don't usually think you are worse off just because someone else becomes better off (the fact that your neighbor won the lottery doesn't reduce your wealth). Still, distress is painful, even when it's irrational or unjustified, and pain is morally import-ant. How much moral weight does their distress have when their lives are no shorter?

The other harm Have-nots suffer is less obvious, though it may lie behind and justify some of the distress, even if they don't quite realize it. This harm concerns the death burden—how bad your death is for you. I'm not talking about the *way* you die—painfully or in your sleep—I'm talking about how much good life you lost by dying. Dying at 20 is worse than dying at 90, all else being equal, for you lose more time when you die at 20. Your death burden was large. Dying at 90 is not so bad; your death burden was relatively small.

What if you're a Have-not in a world where the Haves live far beyond 90—is your death burden at 90 now worse just because more time is avail-able for *other* people? On the one hand, perhaps we should compare your actual life span with other life spans that were possible for *you*. On that measure, your death burden at 90 was not so bad, no matter how distressed you feel. On the other hand, it is technologically possible for you to have extended life, so perhaps we should use an extended life as the measure of how bad your death would be at 90 even if you can't get it. If so, your death at 90 was terrible, and much worse than death at 90 used to be.

Chapter 8 concerns one of the most common objections to developing life extension: only the rich will get it, and we'll end up with a two-caste society of mortals and immortals. This seems pretty clearly unjust. The interesting questions concern what to do about it. We could prevent that injustice by preventing life extension from being developed and used so that no one gets extended life. However, there's a widespread consensus among political philosophers that equality should not be achieved by "leveling-down"—that is, by making better-off people worse off without making anyone else better off (except in a comparative sense). Should we then postpone developing life extension until other, more pressing needs are met, or is that too a form of leveling-down? If the Have-nots deserve access to life extension, must the Will-nots help pay for it, or does that duty fall only on the Haves? What if it's not possible to subsidize life extension for all Have-nots who want it; should we subsidize it for some of them, or should we instead compensate all of them by giving them something we can provide equally to all of them, such as money? (In that event, *no* Have-nots get life extension; is that better than only some of them getting it?) If we can be sure that many Haves will fail to perform their duty to compensate Have-nots, is that a reason to inhibit the development of life extension?

The Haves, Have-nots, and Will-nots have different interests, and life extension will affect their interests in different ways. I argue that the advent of life extension is, on balance, good for the Haves, bad for the Have-nots, and perhaps somewhat bad for the Will-nots. However, the amount of welfare at stake for each Have is probably much greater than the amount of welfare at stake for any given Have-not or Will-not, for each Have has decades or centuries of potential life at stake. We need a framework for thinking about how to adjudicate among their conflicting interests, welfare, and rights and reach an all-things-considered decision about whether life extension should be developed. I tackle that project in chapters 9, 10, and 11.

How should we approach such a project? I'll say more about this in sections 9.1 and 9.2, but here's the short answer: There are those who believe, at least implicitly, that we use ethical theory to settle such questions and that we do this by applying the correct ethical theory to particular questions or cases, much the way legal scholars apply a body of law to a legal question or case. In a simple version of this approach, we would, for example, use the principle of utility, the categorical imperative, or some other fundamental moral principle or principles and ask which resolution of the conflicting interests of Haves, Have-nots, and Will-nots would maximize utility, or could be willed as a universal law, or whatever the fundamental principle tells us to do.

However, there are many ethical theories, including utilitarianism and Kantian ethics, of course, but also Rossian intuitionism, Bernard Gert's moral rules, Thomas Scanlon's contractualism, and some approaches that don't make principles primary at all, such as virtue theory, among others. Moreover, there is no stable consensus on which of these is correct. This suggests that trying to determine the correct moral theory and then applying it will quickly get us bogged down in the vast literature on moral theory long before we can apply any of it. Even if this is an ultimately successful approach, it's not an efficient one nor is it likely to convince anyone who rejects the particular ethical theory I conclude is correct.

Nonetheless, there is a rough consensus on this much: a theory is not plausible unless it's largely consistent with most of our most carefully considered moral judgments about cases and types of cases. Such judgments are often treated as data that moral theories must account for in much the same way a scientific theory must account for empirical data. For example, a theory that doesn't imply that it's usually wrong to deceive others doesn't fit certain data that any plausible moral theory should account for. The same is true for theories that fail to tell us that killing is usually wrong, or that harming others requires exceptional justifications, or that we can disregard promises and agreements at will, or that equality is morally irrelevant. Theories that tell us those things will be rejected *because* they tell us those things. Judgments of this kind about classes of cases are often expressed as what we might call *midlevel moral principles*: do not deceive, do not kill, do not harm others, keep your promises and agreements, treat everyone with equal respect and concern, and so on. In short, we may disagree about fundamental moral principles like the principle of utility or the categorical imperative, but we argue for and against various fundamental principles partly by reference to a set of midlevel principles about which we typically agree.

Therefore, I will not apply a moral theory, and I take no position on which theory is correct. Instead, I will apply some midlevel principles that, I think, are consistent with a wide variety of moral theories. I will use these principles to decide how to adjudicate among the conflicting interests, welfare, and rights of the Haves, Have-nots, and Will-nots. This set of principles contains a principle requiring us to maximize welfare (or utility, if you like) and three other principles that identify some rights: a right to equality, a right to self-determination, and a right against harm. (This is not a complete set of midlevel principles, merely the ones I think we need in order to think about justice and life extension.) When maximizing welfare violates one of those rights, we must respect the right and produce less than maximal welfare, *unless* the welfare to be gained is so great that it outweighs the right.

The fact that I'm referring to maximizing welfare may lead some readers to think that this is a utilitarian approach, but utilitarianism doesn't include rights. Moreover, all ethical theories recognize the importance of welfare, so recognizing it doesn't make us utilitarians. The fact that I'm referring to rights may lead some readers to think that this approach is *in*consistent with utilitarianism, but that isn't true either, for any reasonably sophisticated version of utilitarianism recognizes the existence of seemingly nonconsequentialist but utility-maximizing principles of commonsense morality used in moral deliberation, such as principles of justice or liberty, either through a rule-utilitarian approach or something similar. Again, I'm not committed to any particular ethical theory concerning which more fundamental moral principle explains and justifies the midlevel principles.

The first step in adjudicating among Haves, Have-nots, and Will-nots is to determine which life extension funding policy can be expected to produce the greatest net welfare. In chapter 9, I argue that in the long run, there's more net welfare in an *extended world* (a world with life extension) than in a *nonextended world* (a world without it) once you take into account the extra welfare gained by Haves in their longer lives.

In chapter 10, I discuss whether maximizing welfare in that way violates any rights. There are two rights that *seem* pertinent but are not: the right to equal treatment and the right to self-determination. I will argue that the right to equal treatment is not a reason for or against life extension; that right concerns who gets access to life extension, not whether it should be available at all. I argue that the Have-nots and Will-nots don't suffer violations of their rights of self-determination simply because they find themselves living in a world where Haves extend their own lives. By the same token, the Haves are not denied self-determination if society fails to make life extension available, for the right to self-determination is a negative right against interference, not a positive right of assistance. The rest of chapter 10 focuses on the right not to be harmed and whether Haves are harmed by being prevented from getting a longer life than they have now.

Usually, we may not maximize welfare when doing so violates a right, but there are exceptions when the amount of welfare is great enough and the interest protected by the right is relatively small by comparison. In chapter 11, I develop a decision procedure for deciding when welfare can outweigh a right and use it to decide whether, in an extended world, the welfare of the Haves outweighs the rights-not-to-be-harmed of Will-nots and Have-nots. I conclude that in the short run, the welfare gain for Haves does not outweigh those rights but that it does outweigh those rights in the long run, provided society institutes a policy of Forced Choice to control reproduction by Haves and forestall a Malthusian crisis. This is the crux move in my long argument in defense of life extension.

In chapter 12, I consider bioconservative concerns about human enhancement. Most philosophical discussions of their concerns have construed bioconservatives as saying that enhancement is intrinsically wrong or that anything that is unnatural is wrong. However, they don't say that. I try to present their concerns in a more sympathetic light and in their most plausible versions. In the end, however, most of their arguments do not, by their terms, apply to life extension (even though bioconservatives seem to think they do), and the ones that do are weak, so they do not support Inhibition.

At the end of each chapter from chapter 2 onward, I list the conclusions I believe I've established in that chapter and identify the sections where I argued for them. In chapter 13, I list all these conclusions once again (think of this as an outline of the entire book) and restate the policies I propose concerning (a) whether and how much to fund life extension research, (b) how to respond to the fact that Have-nots lack access to life extension, and (c) what to do about the possibility that making life extension available may eventually bring about a Malthusian crisis.

So the overall structure of the book is this: Chapters 2 through 7 discuss the interests of the three groups, while chapters 8 through 11 concern how to balance those interests. Chapter 12 concerns a kind of objection that doesn't fit easily into those categories. Chapter 13 summarizes the conclusions and policy proposals.

2 The Haves—Would Extended Life Be Boring?

I find myself fascinating.
—Richard Dreyfuss

This too shall pass.
—Middle Eastern proverb

2.1 Do you want to live forever?

Would extended life be a good life? Why not? If it's good to be alive, then it's better to be alive longer. However, as we saw in section 1.9, more than half the people in two surveys said they don't want extended life. This matches what I hear in classrooms and conversations. Perhaps the most common concern is that extended life would be boring. Many survey respondents worry that it might be,[1] and some bioconservatives are sure it would be.[2]

This issue is obvious with radical life extension, where aging is halted or dramatically slowed. Therefore, when I discuss boredom, I will be talking about extended lives that run into centuries or even thousands of years. I won't focus on moderate life extension, where we get merely an extra few decades, only because I'd be surprised if boredom turned out to be a problem in that scenario. If people get bored faster than I think they will, then what I say about radical life extension should carry over to moderate life extension, though probably to a lesser degree.

The best discussion of the boredom issue appears in "The Makropulos Case" by Bernard Williams.[3] Williams discusses immortality, where you cannot die, not extended life, where you slow or halt aging but *can* die. However, his arguments transfer well enough to cases of extended life, at least if those lives are long enough. In fact, the title of his paper comes from a play in which the main character lives for 342 years and finally decides she's seen it all and she's bored with life. She dies by refusing to continue taking the elixir of life that kept her young for three centuries—a case of extended life, not immortality.

I am going to argue that Williams is mistaken and that we should not refuse extended life in order to avoid boredom.

2.2 A dilemma for the very, very old

There are lives worse than death. Is immortality one of them? Williams says immortality would be "intolerable," that "an endless life would be a meaningless one, and that we could have no reason for living eternally a human life." Why? Because endless life becomes endlessly boring, and eventually, we long for death. According to Williams, it's *impossible* for immortal life to avoid becoming intolerably boring.[4] This includes heaven: endless heaven is another hell. (*Hell* is any afterlife that's worse than not having an afterlife.)

Many people read Williams's argument as a dilemma: either you'll suffer intolerable boredom or you won't be immortal. Here is the first horn of the dilemma: If you live long enough, decade after decade, century after century, millennium after millennium, you'll run out of things to do. Eventually, you'll get bored with philosophy, travel, gardening, guitar riffs, improv theater, Ultimate Frisbee, Restoration comedies, tantric sex, and whatever else fills your dance card. You have seen it all and done it all a thousand times, a million times. You can barely stand to exist, and you long for annihilation. You are bored to death without being dead.

Here is the second horn of the dilemma: You might find new interests and values to keep you engaged, but what you find interesting and engaging depends on your character and values. For example, you can't take an interest in competitive sports unless you enjoy competition or an interest in meditation if you have no spiritual side. To acquire a new set of interests to replace the ones that bore you after a few centuries, you need to make some changes to your personality and values. Once you have changed your personality and values so that you can acquire entirely new desires and interests and passions, you avoid boredom—for a while. However, eventually you get tired of the interests that go with the altered version of your personality and values, and you need to alter your personality and values once again so you can acquire another set of new interests.

This may push boredom away indefinitely, but there is a problem: Eventually, after your character and values have changed enough, you have effectively become another person. I've never liked sports, but centuries of ennui and desperation might drive me to stadiums and baseball statistics—at the cost of my very soul. The *you* that existed at the beginning no longer exists; you are now someone else. In effect, you died, and therefore, you were not immortal after all. That's why I said that on the second horn of the dilemma, you will not be immortal.

Of course, you might care about that future (different) self in the same way you care about your descendants in the distant future, but that's a concern for another person, while Williams is interested in whether an extremely long life span is good for *you*. Even if you have reason to care about the different person who will succeed you in the future, this does not mean that endless life is good for you in the self-interested way Williams has in mind.

Why does Williams believe that if you change enough, you eventually become a different person? Williams seems to be assuming that you cannot be the same person over time if you change too much. (He does not quite say this, but I can't make sense of what he says unless I attribute this claim to him.) You can change a lot, but there is a limit, and if you change too much, you become someone new. Imagine that a mad scientist with a strange sense of humor modifies your genes to make them change over time so that over many centuries, you gradually, imperceptibly change from the way you are now until you look just like Pope Francis (even if you were originally female), with his memories and personality, even believing you're the pope, and so on. Even if that change occurs very, very slowly, over many centuries, it seems like you ceased to exist somewhere along the way, only to be replaced with the pope's doppelgänger. (There need not be any sharp line between the earlier person and the later person; becoming a different person can be a matter of degree.) In short, having enough similarity with your earlier self is a necessary condition for personal identity over time, as philosophers put it. You may need more than this to be the same person over time, but if you lack this, you cease to survive over time. I call this the *similarity condition*. Williams appears to assume the similarity condition.

Suppose that by age 940, you have forgotten everything from your first century of life, so you have no memories to connect you with it. (Maybe you have a 900-year-old scrapbook, but that's not the same as remembering those things.) Moreover, you might have entirely different personality traits by age 940. At 40, you were gregarious and optimistic but politically liberal, athletic, had a keen interest in how your children were doing, and loved overseas travel. At 940, you tend to be introspective, you're no longer optimistic (though perhaps not pessimistic either), you have an appreciation for the value of established ways of doing things, you're mildly curious about children you may not have seen for decades, you're just as healthy as you ever were but no longer interested in sports, and you're no longer interested in travel. If the similarity condition is correct, the version of you that existed at age 40 and the version of you that exists at age 940 are not the same person. The two of you are not similar enough. You ceased to exist and were replaced by a different person. Someone exists in what was once your body, but he or she is not you. It may have been a gradual process with no clear line between the first person and the second, but that was the result.

It is important to understand that Williams is not claiming that you *will* change that much over a long enough life. He might agree that you could be the same person over a span of 1,000 years if you're stable enough. He is claiming this: you can avoid boredom *only if* you change that much, in which case you cease to exist. Here is the dilemma again: If you don't change enough, then you don't avoid boredom. If you change enough to avoid boredom, you cease to exist, and if you cease to exist, then you were not "immortal" after all, even if your body exists forever. (In this context, saying that you were not "immortal" is a way of saying that you did not survive over time.) The dilemma, then, is that either immortality is intolerably boring or you aren't immortal.[5] This is what Williams means when he says immortality is necessarily boring—not that boredom can't be avoided, just that you can avoid it only if you cease to survive along the way, not because your body died, but because you changed so much that you were gradually replaced with someone else.

I will treat the first horn of the dilemma as one objection to life extension and the second horn as a separate objection to life extension. In other words, fear of boredom might be one reason not to extend your life, and fear that you will eventually fade away as you evolve into some other person might be another reason not to extend your life. Sections 2.3 through 2.4 are about boredom. Section 2.5 is about personal survival in a very, very long life.

2.3 Can you avoid boredom without fading away?

Would we find extended life boring? We really don't know. This is a factual question, and no one has experienced extended life yet. Moreover, people differ psychologically, so extended life may bore some people and not others.

Scholars of boredom—there's a lively literature on this—have identified three kinds of boredom. First, there is a kind of boredom I'll call *modernity boredom*. This is a mood associated with the rise of a modern, urban, industrial, and technological culture: "a pervasive cultural craving for immediate amusement, risk, and peak sensations, a momentary aesthesis that pulls out of the emptiness and indifference of our everyday lives."[6] The general idea is that we can no longer discern the importance of anything. We have become desensitized to the real significance of things, perhaps because the world is moving so fast that we can't keep up, or perhaps because capitalism reduces everything to commercial value so that real value is overlooked, or perhaps because old religious certainties have been undermined by modern science and it seems like nothing really matters anymore. Is modernity boredom more likely to occur in an extended life? Probably not; this kind of boredom occurs when society changes in certain ways, not simply because we've seen and done it all before. Modernity boredom could occur in an extended life, but it will not occur just because that life is extended.

The second kind of boredom—*chronic boredom*—occurs when you're bored with life itself because nothing engages you. Chronic boredom is associated with withdrawal, anxiety, alienation, alcohol and drug abuse, depression, and suicide.[7] Many psychologists believe that chronic boredom stems from a lack of emotional awareness, an inability to notice one's thoughts, feelings, and moods. One researcher found that patients who suffer from chronic boredom either have no life goals or once had life goals but gave up when the goals were too difficult to attain and attributed their failure to external factors or other people.[8] The key point here is that chronic boredom does not seem to correlate with increasing age. Some young people suffer from it, and some elderly people never have. Therefore, if chronic boredom arises from lack of emotional awareness or giving up on one's life goals, it's not obvious that this would be more likely in an extended life.

People who live extended lives might, however, be likely to suffer from what psychologists call *situational boredom*, which results from repetition or inadequate stimulation.[9] Psychologists say you suffer from situational boredom when you are bored with a particular thing and there's something else you would rather be doing. In other words, you are bored with something in particular, not with all of life in general. On the one hand, this suggests that you will have plenty of time to grow bored with particular things, but that doesn't mean you'll become bored with extended life in general. On the other hand, perhaps you'd become situationally bored with nearly everything after a few hundred years, having exhausted all possible pursuits. In that case your situational boredom might become a kind of chronic boredom, for you no longer have goals to keep you interested in life.

Of course, it's not hard to imagine finding things to do century after century—that's not the problem, at least according to Williams. The trick is to imagine finding new things that don't require radical alterations in your values and personality so that you don't fade into a new person over time. John Martin Fischer suggests that immortality would not be boring if we spend our time on what he calls "repeatable pleasures"—things we just don't get tired of. Because they're repeatable, we don't need to radically alter our personalities and values while developing new interests and pursuits. We might look for new variations on the repeatable pleasures, but we don't need radically new pleasures. Therefore, we can avoid boredom without changing our personalities and values over time and thereby avoid evolving into new people.

Fischer's list of possible repeatable pleasures includes contemplating art and natural beauty, engaging in inquiry and discovery, enjoying good music, good sex, good love, deep friendships, and meditation, among other things.[10] It's a good list, and at least some of those things do seem endlessly repeatable, at least for a few centuries (Fischer's concern is with immortality, which places more stringent demands on the list). After all, nearly everyone has some pleasures at any given age, and they can't all be novel. We still enjoy warm sunshine at 96; we'll probably enjoy it at 796. I have never gotten tired of writing, and I can't imagine I ever will. The repeatable pleasures just have to be varied and mixed up over time.

Williams might object that the pleasures that are endlessly repeatable are not enough for a full life. It's easy to imagine never getting tired of sunshine or meditation, but there may be a limit to how well these things can fill a life that extends into centuries. Making new discoveries seems more promising, but you can't make the same discoveries over and over. Moreover, after a century or two, I might get tired of anything that *looks* like philosophy, even if I've never seen it before.

Fischer might reply that you just need a bigger deck of pleasures to shuffle. Go off and spend time on politics or living as a Buddhist holy man for three or four hundred years, then return to philosophy when you have forgotten all you ever knew about it. You can then discover philosophy as if you were encountering it for the very first time. To do this, you'd have to forget philosophy so thoroughly that you really do read it again as if for the very first time, but that might be possible, given enough time. (If necessary, take a selective amnesia pill to scrub away your philosophical background, like erasing a blackboard for the next class.) Of course, you'd have to retain a lot of other memories, intentions, desires, and beliefs over the long centuries to preserve your identity, while losing the subset that concerned philosophy, and the other memories, etc., would have to be numerous enough for you to count as the same person over time. Still, this may be possible. We can't settle this by speculation from an armchair; it may require living that long and seeing what happens. At the very least, we can't assume that boredom is an insurmountable problem. I am guardedly optimistic.

2.4 The boredom pill

But suppose I'm wrong, and boredom is a harder problem to solve than I think it is. What else can we do about it? A moment ago, I mentioned the possibility of selective amnesia pills. How about pills to prevent boredom and ennui even without amnesia? After all, we have pills for anxiety, ADHD, shopping addiction, and depression. Why not boredom? If we can slow aging, surely we can do this too. (I remember a weekend in college involving peyote buttons, a canyon in Utah, and a twig. I have no idea how long I stared at that twig, but it was fascinating . . . for a very long time. Perhaps a peyote derivative would work.) Along with the drugs and genetic engineering you need to stop aging, perhaps you'll also need a prescription for *Antennui* (anti-ennui).

Williams rejects this solution: "If . . . one tries merely to think away the reaction of boredom, one is no longer supposing an improvement in the circumstances, but merely an impoverishment in his consciousness of them."[11] In other words, if you take the boredom pill, you are not fully conscious of how boring your circumstances really are, and the pill doesn't make them any less boring.

Apparently Williams thinks this is a bad thing ("impoverished" doesn't sound good), but why? It makes you feel better, after all. The comment above is all the explanation he gives us, so we must extrapolate from that. Larry Temkin suggests that Williams rejects the boredom pill in the same spirit that many people reject Robert Nozick's experience machine. Nozick's experience machine gives you simulated experiences, and you can live in there for the rest of your life. Before you go in, you can select whatever experiences give you the most pleasure and happiness (you're unbelievably rich, successful, attractive, and so on), but none of it's real even though it seems completely real—like the pods in *The Matrix*, except that you get to choose the life you'll experience. Would you choose to enter the machine and live in there? Nozick claims you would not. He says that for most of us, being in touch with reality; having real achievements and not imaginary, simulated ones; and actually being the people we think we are, are all more important to us than getting as much pleasure and happiness as possible.[12] Temkin suggests that Williams has a similar concern about the boredom pill: that it's better to be bored when your circumstances really are boring than to feel engaged and interested and falsely believe they are not boring.[13]

So let's ask something like Nozick's question: If you knew your life really *was* boring, would you take a pill that made you feel it was not? Nozick says you would not enter the experience machine because being happy is less important to you than being in touch with reality and having real accomplishments. However, the boredom pill does not separate you from reality the way the experience machine does. Taking the boredom pill doesn't give you the illusion of living an imaginary life or prevent you from having real achievements and real relationships the way the experience machine does. Whatever you do while taking it is really done.

If the pill makes you take an interest in an activity that's actually a waste of time, and something else would be a much better use of your time, then the pill may be keeping you from a better life even if it does not keep you away from real life altogether. (To avoid this, perhaps the pill could be designed to fix on whatever you are doing for the first few days you take the pill so that you "attach" your interest to whatever you consider worthwhile, like a baby duck imprinting on the right mother. If you want to find being a doctor eternally fascinating, then make sure you're working in a medical office while the pill is taking effect.) Anyway, taking the pill does not prevent you from undertaking worthwhile pursuits. If you take the pill after you've lived long enough to have become bored with everything that life has to offer, yet the pill enables you to feel engaged in pursuits that really *are* still worthwhile, then the pill is making your life more meaningful, not less. Any medication can be abused, but that doesn't mean it should never be put to use.

If we think about what it means to have an "impoverishment in [one's] consciousness" when it comes to finding things interesting or boring, we can see another problem with Williams's claim about boredom pills. Williams seems to believe that whether something is boring or interesting is an objective matter, something we might fail to perceive correctly. He would surely grant that there is also a subjective sense of "interesting" and "boring," the affective states we denote that way, but he seems to think that these states are like perceptions. Sometimes they accurately reflect the fact that something is interesting or boring in an objective sense. Sometimes they mislead us, like seeing a mirage. Williams must believe this, for if things were boring or interesting only in a subjective way, there would be no impoverishment of consciousness in thinking they're interesting when they aren't. Moreover, if boredom is only subjective, it's hard to see what's wrong with a boredom pill.

What does it mean for something to be objectively interesting—the opposite of objectively boring? By "interesting," I don't necessarily mean interesting in an intellectual way. I mean engaging in some sense, the sense in which we say you have taken an interest in this or that. Perhaps circumstances are objectively interesting if they are worth taking an interest in, worth caring about, worth trying to preserve or promote in some way. This suggests that being boring has something to do with lacking those qualities. If this is right, then perhaps being objectively interesting is tantamount to being objectively valuable, at least in certain ways, and being objectively boring is tantamount to lacking objective value. How might this apply to a life? We might say that a life is not worth caring about, trying to preserve, and so on if it lacks value. It is an objectively boring life, and if you find it interesting, you have an impoverished consciousness.

However, we tend to think that life, especially conscious human life, has objective value—a kind of intrinsic value. If so, then all human lives are worth taking an interest in, worth preserving, and so on, so they are all objectively interesting and warrant subjective interest. Presumably this will be true of extended lives too. They will have value as long as they continue and therefore warrant subjective interest. They will be objectively interesting even if we don't find them subjectively interesting.

If mere intrinsic moral value is not enough for a life to be objectively interesting, then consider another way a life can have value: You might use your time to make the world a better place. Over the centuries, you work on preserving natural ecosystems, improving the quality of preschool education, or researching a cure for the common cold—and you live long enough to see your efforts succeed. Surely that's enough to make your life worth preserving, promoting, taking an interest in, and so forth. Being interested and engaged in that life—or better yet, in the things you are doing to make the world a better place—seems to justify the very opposite of boredom. This is also true of an extended life, for as long as it continues. It's not objectively boring, provided you use your time properly.

Perhaps Williams believes that there is something wrong with taking pills to fend off or create a particular affective state. If this is what he means, then the problem is not that we find something interesting when it isn't; the problem is that the feeling comes from a pill. However, if what you're doing really is worthwhile but it's wrong to take a pill to adjust your feelings of boredom and engagement so that you see it that way, then we may need to rethink other pills we already take, like antidepressants. If your life is going well but you're depressed, taking an antidepressant helps you see your life the way it really is, and most of us have no objection to achieving that with a pill. These pills restore a proper response to reality. Unless there is some problem with pill-induced perceptions or moods when it comes to boredom that does not arise for other moods and perceptions, then there's probably nothing per se wrong with taking boredom pills.

Moreover, if it's bad to find something interesting when it's not, it should be equally bad to find something boring when it's not boring. We could take corrective pills for both of these conditions: a pill to make us bored when things are objectively boring and a pill to make us engaged when they are objectively interesting. Suppose what you're doing is worth doing, and therefore it's objectively interesting, objectively worthy of your engagement. However, you find your life subjectively boring because you've spent centuries on a succession of good causes or on lots of successful creative projects, and you no longer feel engaged. You have a misperception or, as Williams would put it, an impoverishment of consciousness: you find your life boring when it really isn't. If there is such a thing as objectively justified engagement or boredom, it might well diverge from your subjective sense of engagement or boredom, so a pill for boredom might be a good thing to take. Therefore, taking pills for boredom is not per se wrong.

Finally, it's not clear why the value of accurate perceptions is so great that we should endure any amount of boredom, no matter how painful or prolonged, rather than falsely finding something interesting when it's not, or vice versa. Is a false, drug-induced feeling of engagement in a routine, objectively boring existence really a fate worse than death? Speaking just for myself, the answer is no. If a chemically induced but spurious sense of engagement with life is my best alternative to death, I'll take it. After all, we put up with many things as the price of being alive: serious physical disability, aches and pains, and drugs with unpleasant side effects. Why not this too? Moreover, lots of people seem to live lives of spurious engagement, fascinated by competitive economic display, their local social pecking order, reality TV, or getting stoned at three in the afternoon. Are they better off dead?

2.5 How to survive your survival

Boredom is the first horn of Williams's dilemma. I have argued that boredom is not as big a problem as Williams and many others believe it is. But what about the other horn? Williams says you can avoid intolerable boredom only by changing so much that you eventually turn into someone else and cease to exist. Even if you survive, you fail to survive, so to speak. Thus, even if Williams is wrong to think that we can't avoid boredom without ceasing to exist, it's possible that living long enough will inevitably change you so much that eventually you'll cease to exist anyway, at least if you want to avoid intolerable boredom. After all, Williams doesn't present his dilemma as a choice: it's not as though you get to choose between

(a) immortality with intolerable boredom and (b) avoiding boredom by changing until you mutate into someone else. He just thinks that these exhaust the possibilities. You may or may not have much control over which outcome occurs. Suppose it turns out that you cannot avoid changing so much that eventually you cease to exist, and some future person takes over what is left of your body. Is that an argument against extending your life?

Note that this is not an argument against staying alive as long as you can; it just means that you can't stay alive forever. However, even if this is not a reason against seeking extended life, it may mean that you have less reason to seek it, for your life will not extend as far as you wish. In fairness to Williams, he was not arguing that you have no reason to extend your life; he was arguing that you have no reason to be immortal (a condition where you *cannot* die). However, some people have suggested that our inability to survive enough change over time is a consideration against extended life too. Walter Glannon, for example, doesn't think extended life would be bad for us, but he does think there's no reason to seek it: "The connections between earlier and later mental states in a life of roughly 200 years would be so weak that there would not be any good reason to care about the future selves who had these states. . . . The point is not that virtual immortality would be bad for us as individuals, but that it would not be prudentially desirable."[14] In other words, an extended life that long would not benefit you, so you have no reason to seek it. Is this true?

No. There are three reasons why.

The similarity condition
The first reason is that you might *not* turn into a different person even if you change a great deal over an immensely long lifetime. You might be very different without being a different person.

As I explained in section 2.2, Williams seems to be assuming that personal identity requires a sufficient degree of similarity between the way you are now and the way you are at all other stages of your life, from birth to death. Think of your self at various stages as a succession of *selves*, for a moment. If one of those future selves is too dissimilar from you as you are now, then he or she is a different person, and you came to an end before he or she came into existence (with appropriate provisos for the possibility of a gradual change from one to the next). I called this the *similarity condition*.

Meeting the similarity condition may not be a sufficient condition for two selves being the same person (there might also be other conditions that must be met—for example, they must be part of the same life span in the same body, to put it loosely), but it's supposed to be a necessary condition: if two selves do not meet the similarity condition, they are different people.

Most of us find the similarity condition pretty plausible. We immediately understand the idea that if you change enough over time, you cease to be the same person and become someone else. However, perhaps we shouldn't. Consider what the similarity condition implies. You changed quite profoundly from infancy to adulthood (or even to late childhood) and even more from infancy to old age. The self you were in the first few months of life and the self you will be in your 80s are not very similar at all aside from some shared history and a common genome. And yet we think we're the same person at both ends of that time span. Therefore, even a drastic amount of change may be compatible with being the same person over time. Centuries of life may not change you into a different person.

Imagine yourself at age 11 and again 50 years later, at age 51. Now imagine yourself at age 51 and again 400 years later, at age 451. Will you change more during those 400 years than you did during the 50-year span? During those 50 years, you went through puberty, matured physically, discovered romantic love, acquired a higher education, matured, changed professions and interests and quite possibly your political ideology, saw Paris for the first time, and became a parent and watched your children grow up. What will you do during the next 400 years? Perhaps discover new interests, mature some more, change your political views again, fall in love a few more times, perhaps have a series of spouses over time, and so on. Are those really bigger changes than those you went through (or will) from 11 to 51?

Our reluctance to see our infant selves and our elderly selves as two different people may be why very few theories of personal identity over time contain a similarity condition. Instead, most of them hold that you can change a great deal without losing your identity, provided you change in the right way so that you have *psychological continuity* over time. Your later self is *continuous* with your earlier self (and vice versa) if, at every stage in between, you retain a lot of memories and personality traits from the preceding stage.[15] This is a little like a rope made of overlapping strands, where none of the strands is as long as the rope, but they overlap so that the rope holds together. For example, at 6 you remember some of what you experienced at 3, and at 9 you may have forgotten most of what you experienced at 3, but you remember some of what you remembered at 6, and so on throughout your life. Some theories also require that either your psychological states earlier in life persisted over time or they caused your later psychological states. If they did not persist but did cause later states, then once again you might change drastically over time and still be the same person (according to these theories), provided the causal chain is unbroken.

If theories of personal identity that require psychological continuity are correct, then it's possible for you to survive over centuries or even thousands of years, provided that at each stage, you remember some of what you remembered in the previous stage. For example, at age 4,000 you might be profoundly different from what you were like at age 400, but you are the same person provided that you have continuity and the right causal relation over that entire span.

The Tarzan objection

I believe that psychological continuity theories of personal identity are correct. However, I could be wrong. Let's suppose, just for the sake of argument, that the similarity condition is correct, and some degree of similarity is necessary for survival over time. It still makes sense for you to desire extended life.

The reason is that even if you'll eventually change into a different person (your body outlives you, so to speak), you might desire extended life by stages. As several philosophers have pointed out, even if you now have no reason to care about your distant future self, you do have reason to care about your more immediate future self, who is a lot like you, cares about a lot of the things you do, and *is* you. Moreover, the transition from one self to the next is gradual enough that at any given time, you have reason to care about your self in the near future.[16] For example, at age 50 you don't have reason to care about surviving to age 400, but you do have reason to care about surviving to 100, because at 100, you'll still be you, and at 100, you'll have reason to care about making it to 150 for the same reason, and so on until you're 350 and hoping to make it to 400. I think of this as the *Tarzan objection*: Tarzan can't jump all the way across the jungle, but he can swing as far as the next vine and make his way across the jungle by stages. Even if you change very quickly over time, surely you're the same person from year to year, and in any given year, you have reason to want to survive another year. I consider this objection decisive against the second horn of Williams's dilemma—the claim that we have no reason to desire immortality (or extended life) because we will cease to exist along the way.

However, the Tarzan objection helps us see that if Williams is right about boredom and personal identity, then he is right about one other thing: it does not make sense to desire life extension so that you will live *forever*. Instead, it makes sense to desire life extension so that at any given time, you will want to live a while *longer*. In other words, there are two ways to desire endless life:

1. *Eternity desire:* You now want to be alive at all times in the future (that is, you want to be alive forever).[17]

2. *Ongoing desire:* At each time in the future when you exist, you want to be alive for the time being.[18]

Thus, even if considerations about personal survival over time mean that an eternity desire is not a good reason to seek a very, very long life span, they don't undermine an *ongoing* desire for such a life span.[19] The question is not whether it is desirable for you to live forever. The question is whether it's desirable for you to continue living for a while longer. Of course, if you live long enough, you may indeed become another person, with a radically different personality, character, and desires, and no memory of your earlier self, but that's not a reason to go off your life-extending meds and end it all now, and it won't be a reason to do so then either.[20]

The senility objection

There is another reason why the prospect of changing dramatically over time is not a reason against extending your life. Suppose you change little by little over hundreds of years and gradually fade away, replaced by someone new. (Again, let's grant proponents of identity objections the benefit of the doubt and assume, for the sake of argument, that being the same person over time does require some degree of similarity between your self now and your self then.) If this *is* a way to die, how bad a death is it? Perhaps not that bad; you're physically healthy, so it won't be painful, aside from whatever distress you feel when you think about gradually evolving into a different person and thereby ceasing to exist. How distressing would that be? No more distressing than something we're already familiar with: senility, especially very severe forms of it. There are two ways to think about severe senility: either you change a lot but do not become a different person, or you change so much that you do become a different person. The usual view is that you don't become a different person, but let's consider both alternatives.

Suppose we can survive senility without losing our identity. In that case, if you slowly descend into ever-deeper senility, you lose consciousness without losing your identity, while if you extend your life and change over time without becoming senile, you lose your identity without losing consciousness. Now compare losing consciousness without losing identity (senility) and losing identity without losing consciousness (extended life). When you think about it that way, gradually changing into a different person should be no more distressing than experiencing a gradual descent into senility, and perhaps less so.

Now suppose you can't survive senility without losing your identity (because the change is just too great). In that case, senility is worse than gradually changing into a different person over an extended life, for in senility you not only lose your identity; you lose your faculties too. In extended life, by contrast, you lose your identity but never experience a loss of your faculties; they just cease to be yours, in a sense.

Either way, gradually changing into a new person should be no worse than experiencing a descent into senility, and if this is a death, it's no worse than the death we know now. In fact, it's better. If life extension keeps you alive long enough for you to change that much, then you should not be experiencing much physical decline. Moreover, in the senility scenario, you have less reason to anticipate the immediate future with pleasure, while in the change scenario, you have more reason to anticipate it with pleasure, for the change into a new person is so gradual that you notice it only in hindsight, as something that happened gradually over a long period. From day to day, you never experience that change. Senility, by contrast, settles in far more quickly. The gradual-change scenario is a more attractive way to die than the scenario where you suffer from increasing senility to the point of permanent unconsciousness and, later, biological death.

I would rather fade away by slowly evolving into someone else in a body that lives than fade away into senility in a body that dies. If I do evolve into someone new, it will happen slowly, and I'll be along for the ride, able to enjoy the new horizons it opens up. This strikes me as something to look forward to, not something to dread. I've never been a thrill-seeking extrovert with fundamentalist religious convictions who can't stand being alone, but it might be fun—if not for me, then for him.

2.6 If boredom is unavoidable, is that a reason not to start extended life at all?

Suppose I'm wrong that boredom can be avoided (even by changing your personality and values over time) and that science never, ever develops a boredom pill. If boredom is a real threat, is that a reason to refuse life extension and have a normal life span? Probably not, for two reasons.

The first is simply that boredom takes a while. You may find plenty to keep you engaged for the first 20 years, or even another century or so, even without making any radical changes to your interests and pursuits. It seems needlessly cautious to turn down extended life just because it becomes boring a long time in the future. That's like turning down the last 30 years of your life to avoid dementia and frailty long before they arrive. Why not enjoy some extra time before that happens?

The second reason is that you can discontinue your life extension treatments at any time by ceasing to take drugs or stem cell transplants or by reversing the genetic engineering that keeps you young. This is not suicide in the usual sense; this is resuming aging and living a normally aging life from that point on, just as if you had never extended your life in the first place. (I discuss the ethics of such a suicide in chapter 4.) If extended life becomes unavoidably boring, you can check out.

2.7 Conclusions

This chapter has concerned the Haves—those who extend their lives. Here are my main conclusions in this chapter:

A. Extended life *could* become very boring, but it probably doesn't have to be. You might avoid boredom by acquiring new and repeatable interests and projects over time. (Section 2.3) You might also avoid boredom by taking a pill, much as we now use pills to avoid depression, anxiety, and other states we do not like. There's nothing wrong with taking boredom pills. (Section 2.4)

B. Even if you change so much over time that eventually you become a different person, either because you acquired new interests and values or because you just lived long enough, this is not a reason to turn down life extension. It's rational to have an *ongoing desire*, a desire at any given time to continue existing a while longer. Moreover, evolving into a new person is at least no worse than descending into senility and moving closer to death, and probably better. (Section 2.5)

C. Even if extended life is unavoidably boring, that's not a good reason to refuse extended life. At most, it's a reason to discontinue extended life after it has become unavoidably boring. (Section 2.6)

3   The Haves—Death Benefits and the Human Condition

3.1 Introduction

Is extended life a waste of time? Some writers say it will sap our motivation to use our time well and accomplish something. We'll have so much time that time itself will be less valuable (a thing cannot have value when we have an infinite supply of it), and what we do with that time will lack value too. Nothing will matter, and extended life will lack value and meaning. Moreover, it will be harder, if not impossible, to develop certain virtues and other aspects of good moral character, for character develops under constraints, and extended life is unconstrained—it just goes on and on. Some of the writers who make these arguments speak of the "benefits" of death.

I'll argue that the death benefits are either nonexistent or small enough to be easily outweighed by the benefits of extended life. However, although bioconservatives are wrong to think that extended life is a change for the worse, they are right to think extended life is not just a longer version of life as we know it. Life extension presents a new kind of human condition and a new relationship with death—and not just in the obvious sense of postponing death indefinitely.[1] Apart from being longer, the new human condition is not obviously better or worse—just different.

3.2 Making a case for extended life

Possible drawbacks to extended life have dominated the literature on life extension. In my view, they have received far too much attention, not just because the drawbacks are insignificant, but also because focusing so heavily on possible drawbacks sets up a lopsided discussion. I suspect that this happens in part because it's easier to make interesting arguments that extended life is bad than it is to make interesting arguments that extended life is desirable. What can we say, really, except that extended life gives us more time for all the good things a life can contain? There are many more arguments against extended life than for it, and the sheer number (and the hydra-headed variety of ways to state them) can make it seem like there's a fire somewhere under all that smoke. However, focusing excessively on these concerns can lead us to overestimate the downside to extended life and undervalue the upside.

So let me try to forestall a lopsided discussion by briefly summarizing the case for extended life. Extended life gives you more time to accomplish your life goals and contribute to the world. You would live long enough to see what the future brings and to benefit from the marvels of a more advanced science. You would live alongside the posterity you already care about. A longer life span might give you greater wisdom and maturity, and if enough of us undergo this, the world might be less violent, less prone to war. We might value the future more and take a long-range view when it comes to the environment and the welfare of people generations in the future. Finally, and much closer to our present concerns, life extension research may be the most effective way to combat age-related diseases—halt aging and you reduce the incidence of cancer, heart disease, stroke, and diabetes and the suffering they cause.

Consider one possible extended life. Our heroine used life-extending technology at 23, effectively halting aging in her mid-20s. She finished grad school by her early 30s, earning advanced degrees in genetics and public policy, then began a successful career at the National Institutes of Health, where she helped draft legislation on gene therapy and health insurance. As the years went by, she married, raised children, and took sabbaticals every few years to travel on all five continents. She learned languages, experimented with new spiritual perspectives, published a volume of poems, twice ran for Congress, and served on her city council. In her 70s—while still looking and feeling young—she separated from her husband and entered a Buddhist convent for three years of meditation, where she developed a talent for ink-brush painting and took long hikes in the Japanese Alps. In her 90s, she took advantage of the newer enhancements made possible by applied neuroscience and achieved a higher IQ, a better memory, and easier access to altered states of consciousness through meditation.

Using these enhancements she was able to reach new heights of meditative awareness and altered consciousness and then returned to university to study art history and religion. She again ran for Congress with no more luck than before. After a century or so, she made virtual-reality trips to a variety of undersea locales, other planets, and fictional worlds, living in some of them for months or years on end. She emerged into reality from time to time to check up on her investments. She entered into a variety of relationships with various genders at various times and experimented with polyamory. She lived long enough to see the world settle down and become largely free of poverty, war, and short-sighted environmental abuse. She is 400 years old and shows no visible signs of aging.

In that life, everything went right (aside from three fruitless runs at Congress). It could also, of course, go *this* way: She didn't get into the grad school of her choice, had difficulty getting grants and getting published, lost decades in an abusive relationship that went nowhere, suffered the death of a child, broke her knee on a hike in the Japanese Alps, and the critics panned her book of poetry. The point is not that extended life is a blissful march from one triumph to the next or a living hell. It can go as well or as badly as any other life, and over enough time, it may well do both. The point is that it's as promising as any other life, so it's probably in your interest to try it out for a while. If it doesn't work out, you can always terminate your life extension treatments and resume aging, so what have you got to lose by giving it a try?

According to some writers, quite a lot.

3.3 General problems with bioconservative arguments

These writers are often called "bioconservatives," for they oppose altering human nature and wish to conserve it. We'll meet them later in this chapter (I cover them at greater length in chapter 12). Bioconservatives think we are better off without extended life and that a normal life span is much better for us. Before I present their arguments, let me note four general problems with their arguments.

First, *extended life is not immortality*. As we'll see, some of these arguments were originally proposed in connection with immortality, where dying is impossible. They are meant to show that being immortal is not desirable. Whether or not that's true, arguments about immortal life don't transfer well to extended life, where we're mortal and will die sooner or later from some cause of death unrelated to aging even if we don't age at all. Death's power to motivate us, force us to invest in things bigger than ourselves, develop courage, avoid boredom, appreciate the transient beauties of life, and so on may be muted, but in an extended life, it's never gone.

Second, if we're going to talk about immortality, then we should extend these points to all forms of immortality. If we do, however, we get some strange results. God (if he exists) is immortal; is he deficient in virtue? Does he procrastinate, lack motivation, exhibit narcissism and selfishness, or suffer boredom? Does his life lack meaning? But God is different; he's perfect and responds differently to immortality than we would. Very well, let's switch to human immortality: Suppose there's an afterlife where we exist as immortals (a possibility I accept). Do we lose our virtue once we get there? Do we suffer boredom, procrastinate, become more selfish, fail to appreciate life, or find the afterlife meaningless? It is hard to say until we get there, but ask yourself this: Suppose you know no more about the afterlife than you do now, and you are given a choice: you can (a) have an afterlife (without knowing whether you will like it) or (b) cease to exist at the moment of death. Hesitation and anxiety are understandable here, but would you really give up an afterlife just to play it safe?

Third, we get odd results if we run these arguments backward. Suppose, for discussion, that these things are true of extended life. What, then, would happen if we had a significantly shorter life span? Imagine that a mutated virus gets loose and rewrites the DNA of every human on earth in a way that accelerates aging so that we naturally age in our 30s and die in our 40s. (If it helps, imagine that this happened tens of thousands of years ago, so we've long since adjusted to this.) Would we then be even *more* motivated to achieve something, less inclined to waste our years, more appreciative of what life has to offer than we are in a normal life span? Would we be more virtuous and have better moral character? Would we be less given to narcissism and selfishness than we are now? Do people who die early from disease, knowing in advance this will happen, have more meaningful lives than the rest of us? In short, if we naturally had 45-year life spans, would we have better lives and be better people? Probably not.

Fourth, pointing out drawbacks to extended life does not establish that there are no advantages to it or tell us whether the drawbacks outweigh the advantages. However, bioconservative writers focus only on the downside.

This is the lopsided discussion I warned you about. We don't make other important choices by focusing only the downside. You wouldn't decide which house to buy just by looking at what's wrong: a noisy street, tiny closets, or a roof that needs replacing. We evaluate other things by considering the whole picture, cons *and* pros, and we should do the same when it comes to extended life.

With that in mind, let's consider the supposed drawbacks the bioconservatives warn against.

3.4 Accepting death

Leon Kass says having a normal life span with a scheduled death makes it easier to accept the fact that you will die one day.[2] He may be right. Death may be harder to accept when it arrives randomly and there's no stage of physical decline during which you can console yourself by thinking, "Nearly everyone dies by this stage, so I've done well to make it this far." Death will be less feared as imminent but perhaps more feared as unexpected. Moreover, it will deprive of us far more time than it does now. All of this may well make death harder to accept.

This is possible. It's also possible that if you live long enough, you may feel that you have accomplished and experienced enough for one life, that death no longer threatens your ability to have a full life, and thereby become *less* fearful of death. Moreover, this is another argument we don't want to run backward: you would not shorten your life in order to make death even easier to accept, so it makes little sense to refuse to extend it for that reason. Refusing life extension to make death easier to accept is a bit like driving around without a seat belt so you can more easily accept the risk of becoming mangled in an accident. Being less prepared for death is a drawback we can live with.

However, I will concede a related drawback. I call it *lucky streak anxiety*. Right now we don't worry very much about dying from non-age-related causes, for we're more likely to die of age-related causes. As we get older, we have less time left, so an accidental death would not deprive us of very much time anyway. If, however, we die only from causes that are not age-related, and if death deprives us of centuries of life, then we might become far more anxious about this than we are now. It might become harder to go skiing, drive a car, or leave the house for fear of accidents.[3] As the years go by, you might become more and more anxious about the unscheduled moment of your death, wondering how long your lucky streak will last. I've been told that some optimists who believe they'll get life extension in the future travel around in Hummers: they don't want to die in an accident before they can get life extension. They got the anxiety before they got the luck.

3.5 Motivation and procrastination

Some bioconservatives argue that having a normal life span with a scheduled death forces us to use our time well, make our limited years count, accomplish something, and take life seriously.[4] Hans Jonas says, "Perhaps a nonnegotiable limit to our expected time is necessary for each of us as the incentive to number our days and make them count." Kass claims that "to know and feel that one goes around only once, and that the deadline is not out of sight, is for many people the necessary spur to the pursuit of something worthwhile." In an extended life you'll dawdle, like a student with an extension who never finishes that term paper, and end up living a frivolous, aimless life.[5] Something like this thought may lie behind Psalm 90, which asks God to "teach us to number our days so that we may get a heart of wisdom." You're more likely to make wise choices if you keep track of your remaining time so that you don't waste it, and you can do that more easily if you know how much time you have left. If you extend your life, you'll procrastinate on an epic scale.

Yes, you might waste time. Most of us already do. Then again, you might eventually get tired of wasting time and start using it well, or you might use each year less well but get more done overall because you have so many more years to work with. Or you might come to believe that accomplishing a lot is overrated and that the point of life is not to fill out a resume. Or you might be spurred to action by the knowledge that death is unscheduled and can come any time you cross the street. Besides, this drawback will not materialize in all lives to the same extent; much depends on your values and your drive. Finally, this is less a threat in moderately extended lives, where aging is merely slowed somewhat and you live for, say, a mere 150 years or so. We simply don't know how big a problem procrastination might be, but there is reason to think it's a problem we can handle. Turning down extended life in order to avoid procrastination is like avoiding college so that you never turn in a late assignment.

The procrastination concern commits every one of the four mistakes I attributed to bioconservatives in the previous section. First, even if this is an issue in an immortal life, extended life is not immortality. Second, this argument suggests that beings who are immortal (such as us, in an afterlife) will fail to get things done, and thus an afterlife is undesirable. Third, this is another argument we don't want to run backward; we don't want to shorten our lives in order to be more productive. Finally, this argument overlooks the advantages of having more time when it comes to getting things done.

3.6 The meaning of life

Some thinkers claim that an endless life lacks the things that make life meaningful. In its extreme version, this amounts to saying that aging and a normal life span are necessary for the meaning of life. Kass asks, "Could life be serious or meaningful without the limit of mortality? Is not the limit on our time the ground of our taking life seriously and living it passionately?"[6] Kass could be saying that living under the death sentence of normal aging motivates us to use our time well—a version of the procrastination objection. However, he might also mean what John Pauley says: "Our finitude is so deeply rooted in our being that it necessarily structures the meaningfulness of our lives. On the condition of an infinite amount of time in existence, nothing would matter for persons."[7] Nothing would matter, and extended life would lack value and meaning.

The meaning of life is a complex topic, but roughly speaking, to say that a life has meaning is to say that it is worth living—that it has value for the person who is living it. There is more to it than that, but for our purposes, we don't need to define or characterize the kind of value necessary for a meaningful life. It will be enough to list some of the things we commonly think can make a life meaningful: creative activity, raising children and forming a family, intellectual development, working to make the world a better place, helping those in need, developing your talents, and overcoming challenges to gain something worthwhile, to name just a few.

With that in mind, perhaps Kass and Pauley are saying that most things become less valuable the more of them we have and that this is also true of the things that make life meaningful. We value silver partly because it's scarce, and we would value great paintings less if every painter had the

talent of a Picasso or a Rembrandt. The morning's first cup of coffee is usually the best. Therefore, the activities, experiences, and accomplishments
that have value and make life meaningful will have less value the more
of them we have. The longer we live, the more of them we'll have, and
eventually we'll have so many that they no longer have enough value to
make our lives meaningful. For example, writing philosophy papers gives
some meaning to my life, but it's hard to imagine that they would provide
as much meaning after I've written hundreds of them over a century or
two. That, anyway, is what I think Kass and Pauley are suggesting.

To evaluate this argument we must distinguish two ways in which
life might have meaning. First, it might have *objective meaning*. That is,
a life might contain things that really do have value apart from whether
or not we think they do. Second, a life might have *subjective meaning*. A
life has subjective meaning when *we* find it meaningful; in other words,
when we feel and believe that the things it contains have objective value.
They engage and satisfy us, and we consider them worth doing or having. Obviously a life might have subjective meaning while lacking objective meaning. You might, for example, value fame and social standing far
too highly and overlook other things of real value. A life might also have
objective meaning without having subjective meaning. Perhaps you don't
think the objectively meaningful things in your life have value, or perhaps
you do but you take no satisfaction in achieving or producing them. I'm
inclined to say that a life can't have full objective value unless it has subjective value as well. If you don't enjoy or feel engaged by your life, then
your life lacks at least one thing that's objectively valuable: the quality of
engaging and satisfying you.

With the objective/subjective distinction in mind, we can distinguish
two questions about the meaning of extended life. First, does an extended
life have less objective meaning over time? Second, does an extended life
have less subjective meaning over time? As for objective meaning, it's hard
to see how creative work, making the world a better place, conserving and
appreciating works of art, raising children and caring for friends, developing your talents, or anything else that might give a life objective value
will be less valuable just because there is more of it over time. After all,
the objective value of rearing your children is not diminished by the fact
that so many other people are rearing children all around you or by the
fact that thousands of generations have reared children before you. Making

the world a better place does not become less valuable over time unless it becomes harder, over time, to achieve any further noticeable improvements because so much has been improved already, somewhat like trying to improve train service in Switzerland. If something of value does not become less valuable just because so many others are doing it and so many others have done it before you, then it won't become objectively less valuable just because *you* have been doing it for hundreds of years. So far as objective meaning is concerned, extended life is no threat to the meaning of life.

Subjective meaning is another matter, and I suspect this is what Kass and Pauley are concerned about. Rearing children might always be highly meaningful in an objective sense, but it's understandable that you might find it less satisfying and engaging after you've reared 137 of them over the last thousand years. It's conceivable that even a great composer might grow bored with composing symphonies after 800 years, even if the quality and variety of his output never falters. It's certainly not inevitable that you or he might grow tired of children or music, but it's also certainly possible. Extended life might come to lack subjective meaning.

But notice that we've seen this issue before. This is the boredom issue in another guise. To say that you no longer find something engaging, satisfying, enjoyable, or meaningful, is a way of saying you're bored with it. Conversely, to say that extended life might become boring is a way to say it might lose subjective meaning over time. When we discussed boredom, we were also discussing subjective meaning, and what I said about it in chapter 2 applies equally well here.

3.7 Character and virtue

As we all know, we develop some aspects of character—virtues, if you will—partly by responding to challenges or experiences of various kinds. We develop courage by facing our fears, patience by enduring something we want to avoid, and compassion and empathy by coming to understand other people better. Some virtues involve dealing with (among other things) the risk or inevitability of death. For this reason, some thinkers believe that it would be harder to develop those virtues in an immortal life, where death is not even possible, let alone inevitable. Some writers have suggested that this is also true of extended life. Stephen G. Post says, "It is impossible to imagine our capacities for kindness and benevolence evolving without a dominating investment in the young rather than in ourselves."[8] Bill McKibben believes that "without the possibility of death, heroism would disappear."[9] Leon Kass believes that life extension makes it harder to acquire virtue:

> To be mortal means that it is possible to give one's life, not only in one moment, say, on the field of battle, but also in the many other ways in which we are able in action to rise above attachment to survival. . . . We free ourselves from fear, from bodily pleasures, or from attachments to wealth—all largely connected with survival—and in doing virtuous deeds overcome the weight of our neediness; yet for this nobility, vulnerability and mortality are the necessary conditions. The immortals cannot be noble.[10]

Notice the reference to immortality in an argument about extended life.

Martha Nussbaum has made the same argument at greater length, and although she limits her remarks to immortality (I won't charge her with trying to extend it to extended life), we can see the argument more clearly in her version. She says that immortals cannot develop certain virtues, or at least not to the same extent that we do, for those virtues can't be fully developed without facing danger and making sacrifices, especially in the face of death. There would be no virtue of courage, for "courage consists in a certain way of acting and reacting in the face of death and the risk of death. A being who cannot take that risk cannot have that virtue—or can have, as we in fact see with the gods and their attitude to pain, only a pale simulacrum of it."[11] According to Nussbaum, "That component of friendship, love, and love of country that consists in a willingness to give up one's life for the other must be absent as well," for as we see in Homer, "there is a kind of laxness and lightness in the relationships of the gods, a kind of playful unheroic quality that contrasts sharply with the more intense character of human love and friendship, and has, clearly, a different sort of value."[12] There would also be no virtue of moderation, for "moderation, as we know it, is a management of appetite in a being for whom excesses of certain sorts can bring illness and eventually death." The same is true of the virtues of justice and generosity, for "political justice and private generosity are concerned with the allocation of resources like food, seen as necessary for life itself." Without the threat of death, "distribution would not matter, or not matter in the same way and to the same extent."[13]

Is this a sound argument? Let's start by considering it in the context of immortality. Even there, this argument is weak. You can acquire virtues that require facing serious dangers and challenges even if you cannot die.[14] Courage, for example, can be developed in the face of dangers that are not lethal. Many things would still be very difficult even in an endless, immortal life, such as making a marriage work (even harder, probably), writing good philosophy papers, learning to live without things you cannot get, and trying to surpass your previous achievements (or accepting the fact that you can't).[15] Some choices are still more important than others, and it matters when we make them. Some things can be achieved only once, and some have permanent consequences. Some goals are very difficult to achieve, and some may be impossible. Boredom itself, if it turns out to be a problem, could be one of the obstacles we must struggle to overcome or deal with.[16]

The argument is even weaker when it comes to extended life. Even for those who never age, death is still a threat on the battlefield, or while treating plague victims, or when rescuing children from a burning house. Most of the situations where you develop virtue by facing or risking death involve potential causes of death that are not age-related, so extending life will not affect their acquisition. Moreover, the virtues needed to deal with resource shortages and distribution—moderation, justice, and generosity—are just as important in an extended life. After all, famine is just as lethal for those who never age, and developing techniques of life extension does not guarantee we will have enough resources for everyone. The versions of this argument that are deployed against life extension exhibit three of the argumentative mistakes I mentioned earlier: extended life is not immortality, it is not clear that the argument works even for immortality, and we don't want to run the argument backward (I'll let you work it out this time).

In fairness, Nussbaum qualified her argument in ways that indicate she would not apply it to life extension. She says that "the closer we come to reimporting mortality—for example, by allowing the possibility of permanent unbearable pain, or crippling handicaps—the closer we come to a human sense of the virtues and their importance." She also doesn't claim that an immortal life would have no value—or even less value—just that we cannot imagine the kind of value an immortal life would have: "There would be other sources of value, no doubt, within such an existence. But its constitutive conditions would be so entirely different from ours that we cannot really imagine what they would be." Finally, she admits that "the argument is incomplete. For it should, ultimately, investigate not only death, but other limits as well: the human being's susceptibility to pain and disease, our need for food and drink, our proneness to accidents of various kinds, our birth into the world as vulnerable babies."[17] In short, it's a mistake to extend Nussbaum's argument to extended life, but it's not Nussbaum's mistake.

3.8 Narcissism and transcendence

Kass has another concern about extended life and the development of character. He worries that we might become more selfish, more narcissistic, and less able to transcend our own lives by identifying with and caring about other people and posterity: "Simply to covet a prolonged life span for ourselves is both a sign and a cause of our failure to open ourselves to procreation and to any higher purpose. . . . For the desire to prolong youthfulness is not only a childish desire to eat one's life and keep it; it is also an expression of a childish and narcissistic wish incompatible with devotion to posterity."[18]

This argument probably doesn't work for immortality, and it certainly doesn't work for extended life. Suppose you live for centuries. You can still be concerned with the children you produced centuries ago (assuming they still show up for Christmas). You'll still have a posterity, it's just that you'll live to see more of it (this may tend to increase, rather than undermine, your attachment to posterity). Even if your interest in those people wanes after 150 years, you might grow attached to other people along the way. As for causes, you can take them on even if you never age: living longer does not obviously undermine your motivation to preserve the natural environment, eradicate poverty, or promote the arts. Now run the argument backward: nature has already shown us where that leads; so far as I know, there is no evidence that people who know they will die decades ahead of schedule, such as victims of Huntington's disease, are less selfish or more concerned with posterity than anyone else. Therefore, we have no evidence to believe that extending life will make us more selfish or less concerned with posterity.

But suppose, for the sake of argument, that the bioconservatives are right—that extended life will prevent us from developing virtue and character as fully as we would in a normal life span. Is that a reason to turn down extended life? I believe not. Even if I'm wrong and our moral character would suffer in an extended life, it's not clear that we should refuse extended life. I'm not talking about the "should" of self-interest but about the moral value of life and the moral value of virtue. In other words, perhaps we should choose extended life even if that meant we would have less well-developed moral characters. (Well, maybe not if it meant becoming amoral and corrupt, but even arch bioconservatives don't expect things to get that bad.) Many of us think that human life has intrinsic moral value and that every life is morally important. If so, then ending it sooner than we need to seems wrong, for ending it cancels something that has moral value. All else being equal, a more virtuous life probably has greater moral value than a less virtuous life, but giving up centuries of that life just to make it more virtuous is quite a trade-off. The total moral value of a much longer life with a somewhat lower level of virtue may exceed the total moral value of a much shorter life with a somewhat higher level of virtue. The gain in virtue would have to be staggering for that decision to make any sense.

Of course, it seems odd to say that the right thing to do is to be less virtuous in order to extend your life. But maybe that's not so odd after all. Both your life and your interests have moral weight too, and when your life and interests are sufficiently threatened by the trade-offs needed to attain some high degree of virtue, perhaps the moral value of that life and those interests can outweigh the moral value of that extra margin of virtue. In that case, the moral value of that extra virtue is outweighed by the moral value of extra years of life.

3.9 Adaptive preferences (sour grapes)

In this chapter and chapter 2, we considered several alleged drawbacks to extended life: boredom, loss of personal identity over time, procrastination, time losing its value, and various difficulties in developing moral character. I argued that these drawbacks are either nonexistent or not serious enough to outweigh the value of additional life. We might summarize the point this way: Suppose that slowing or halting aging means that your existence would be characterized by all the alleged drawbacks—is that life really worse than death? It's hard to understand how it could be, given other choices we already make. We put up with quite a lot before we decide death is preferable: serious chronic pain, extreme disability, loss of loved ones, loss of dignity, and other forms of suffering. Why, then, do so many people say they would turn down life extension if it were offered to them? Are we *that* concerned about boredom and our moral character?

Judging from my conversations with many people about life extension, some of this may stem from confusing life extension with conventional end-of-life care, with its feeding tubes, dementia, and bed sores. No one wants decades of life in that condition. For many others, much of their diffidence about extended life might be chalked up to the extreme novelty of slowing or halting aging—to instinctive personal conservatism. Most of us haven't thought very much about this, and it's easy to cling to the familiar. After all, life without aging is a far more radical modification of the human condition than we find even in most science fiction. Science fiction gives us starships that travel faster than light and machines that travel back in time but rarely gives us humans who do not age. It's natural to hesitate when confronted with profound change.

However, as I mentioned in chapter 1, a lot of this is probably *adaptive preference formation*. The fox in Aesop's fable adapted what he wanted to what he could get and convinced himself the grapes were not worth having. He formed an adaptive preference.[19] Adaptive preferences may result from social conditioning and indoctrination, learning and experience, and false expectations, among other things.[20] Whatever their cause, they seem to be a kind of unconscious psychological adjustment that reduces tension and frustration.[21] This can be beneficial. Rejecting something we can't get may be in our interest when we really can't get it, for that makes us feel better about not having it. However, this can also be very dangerous, especially when we really can get the grapes—or at least have a legitimate right to them. For example, women subject to oppression and abuse often come to see such treatment as normal and to feel that they aren't entitled to anything better.[22] In situations like that, adaptive preferences are destructive and get in the way of justice.

Adaptive preferences cause trouble when we think about life extension. We want to know whether we would want life extension if we could get it. We can't learn that by consulting attitudes that help us accept a normal life span, where we cannot avoid aging. Even now most of us (me included) will not live long enough to get life extension or won't be able to afford it. In light of that, it's more comfortable to believe life extension is not worth having than to believe we are missing out on something big. Unfortunately, that preference is a poor guide when we're thinking about whether extended life is desirable for those who *can* get it. Even if it's in our interest to believe that extended life is not worth having, we can't think clearly about the merits of life extension research and development unless we undertake the painful task of setting those preferences aside. When we ask whether we would accept extended life if it were offered to us, we need to think and feel as if it really were an option, however uncomfortable that is to imagine.

3.10 Unscheduled death and the new human condition

In this section I want to set aside what critics of life extension think extended life would be like and offer my own view of this. Although I reject the bioconservative claims about extended life, they are right to think that it's not just a longer version of life as we know it. Extended life changes our relationship with death in four ways and presents a new version of the human condition. The first two ways appear both in moderately extended lives, where we might live a few more decades at best, and in radically extended lives, where we might live for centuries. The third and fourth features appear only in radically extended lives:

a. Even if you decline life extension and age normally, aging is no longer inevitable—it is elective.

b. Extended life is optional at any time, for you can always cease your life-extending treatments and resume aging.

c. Death will be unscheduled.

d. Your life expectancy will always be the same, and death will deprive you of more time than it does in a normal life span.

Aging will be elective
Aging and dying from predictable, age-related causes have always been unavoidable, unalterable aspects of the human condition. However, for those who can afford life extension, normal aging and a normal life span will be elective, a matter of choice. This is true for the Will-nots as well as the Haves, and it's true in both moderately and radically extended lives.

You can always reverse the decision to extend your life and resume aging
As I've pointed out before, the decision to slow or halt your aging will be a reversible choice. (This is an aspect of the fact that it is elective.) You

can cease taking the necessary drugs or stem cell transplants or reverse the genetic engineering back to its natural state. If extending your life were an irreversible decision, you might want to play it safe and not take a chance on it, but because it's reversible, you can give it a try and find out for yourself. Some questions should not be answered from armchairs.

Death will be unscheduled

The next feature of the new human condition arises mainly in radically extended lives. For this one, assume that life extension completely prevents, halts, or reverses aging so that people die only from causes that are not related to aging, such as violence, accidents, or disease. (The feature I'm going to discuss is also present in moderately extended lives, but only to a lesser extent, so I'll focus on radically extended lives.)

As I said, life extension is not immortality. If you're immortal, you cannot die. If you merely halt aging, you will die sooner or later, but not from age-related causes. You'll die from something like an accident, violence, disease, or suicidal behavior. It's impossible to avoid such things forever. This is obvious, but it's worth mentioning, for as we saw, many writers who argue that extended life is bad for you use arguments originally developed to show that immortality is bad for you. Those arguments don't transfer well to extended life.

However, because you'll die only from causes that don't correlate with age, death will not be more likely the older you get (at least in lives where we never age at all). As things now stand, even the young can foresee that death will be imminent for them in a few decades, and the elderly know it could come at any time. We have a sense of how much time is left. However, in radically extended life, the odds of death are the same at any time, and they never go up. Therefore, you will never know—even within very rough limits—how much time you have left. Death will be *unscheduled* and (assuming you haven't caught a lethal disease or gone into battle) never imminent. Your life span will not be infinite (that's immortality), but it will be *indefinite*. You will not count the years until death. You will count the years since birth.

No matter how old you get, your life expectancy in an extended life is always the same: several centuries or more

Here is another feature that arises only in radically extended lives; it's closely related to the fact that death is unscheduled. In an extended life, your life expectancy will always be the same. Because death is not more likely the older you are, your remaining expected years of life will always amount to several centuries, whether you're 21 or 921. No matter how old you are, death will always deprive you of the same vast span of time, measured in centuries.

You might think that the odds go up over time even if there are no age-related causes of death. After all, how long can you avoid an accident on the highways? Surely your luck is running out after few centuries, so death *must* be more likely at age 800 than it was at 200. If you think about extended life that way, then you might think that having a life expectancy of 1,000 years (as you would if you never aged) means you will live until you are around 1,000 and then die, barring an untimely death earlier in life—or at least that death becomes more likely as you approach 1,000.

However, thinking this way is a mistake; it's an example of the gambler's fallacy. Suppose you have a coin that's evenly balanced (a "fair coin"), someone tosses it, and the person tossing it is not manipulating it to make one side more likely to turn up than the other (she's a "fair tosser"). In other words, imagine that the outcome of each coin toss is independent of the outcome of any other coin toss and that the odds of landing heads is 1 out of 2. Now suppose the coin lands heads four times in a row. Most of us think that because landing heads five times in a row is so unlikely, the odds that it will now land tails are better than 1 out of 2. However, that is false. The odds of landing heads are always 1 out of 2, assuming a fair coin and a fair tosser. The odds of the coin landing heads on the fifth toss are 1/2, and thus the odds of landing heads five times in a row (after the first four tosses) are 1/2, even though, at the outset, the odds of landing heads five times in a row is 1 out of 32.

Now consider life extension. Let's say that death by disease, violence, or accidents are all accidents in a more general sense. Let's also assume that the odds of such accidents do not vary from year to year (you're a "fair coin") and that you're not doing anything to change those odds from year to year (you're a "fair tosser"). If these conditions obtain, then your odds of death in any given year are always the same. If you live to be 1,000, you are just as likely to live to 1,001 as you were to live to 51 when you turned 50. It's true that your odds of making it to a later age are always less than the odds of your making it to some earlier age; for example, your odds of making it from 900 to 950 are much better than your odds of making it from

900 to 1,800, but if you make it to 1,799, your odds at 1,799 of making it to 1,800 are the same as your odds at 900 of making it to 901. The average extended life span of 1,000 years is simply the average of individual life spans that have no set limit, determined by the accident rate and similar factors.

This means that no matter how old you are when you die, death deprives you of roughly 1,000 years (the life expectancy we have if we never age and we die only from causes unrelated to aging)—even if you are 1,000 when you die. Thus, your life expectancy in an extended life is always longer than it is in a nonextended life, all else being equal.

This may seem to make your death worse, for we tend to think that death is worse the more time you lose when you die. That's why it's worse to die at 20 than at 80; at 20, you lose 60 years, and at 80, you lose much less. If you completely halt aging, then death will deprive you of centuries—far more than we lose now. Thus, in one sense, life extension makes death worse. However, as Jeff McMahan has argued, when we measure how bad a death is, we not only consider how many years the person lost by dying at a given age but also consider how much that person gained while he or she was alive. As McMahan puts it, we discount the years we lose in death by the years we gained before dying.[23] For example, a patient with progeria (a disease that mimics accelerated aging and kills people in their teens) and an elderly patient may both have only a year or two left to live. If we consider only the time lost, their deaths look equally bad. However, most of us think the progeria victim's death is worse, so we must be considering something else too. McMahan says the progeria patient's death is worse because it's discounted by only 15 years of life, while the elderly patient's death is better because it's discounted by 70 years of life. The elderly patient gained far more during her life than the progeria victim gained during his simply because her life was so much longer (all else being equal).

Therefore, even though death deprives you of much more when you're not aging, your death neither seems nor is as tragic as that vast deprivation would suggest. Suppose, for example, that you have life extension and die at 200. If the life expectancy estimate of 1,000 years is correct, and your remaining life expectancy is always 1,000 years, then you lost 1,000 years but gained 200 years. Now compare that with a normally aging person who dies at 20 when he could have lived to 100. The ratio—2 to 10—is the same in both cases, and the total deprivation is greater in the case of the extended life, yet my judgment is that the death of the 20-year-old is worse than the death of the 200-year-old. That suggests that not only do we discount a deprivation by previous gains, but we also weight the years gained much more heavily than the years lost. The 200 years gained in the extended life outweighs the 80 years lost in the nonextended case, but those 200 years also outweigh the 1,000 years lost in the extended life so decisively that we're better off living an extended life even if death deprives us of so much more. Therefore, life extension does not make your death worse, even though you lose centuries when you die.

Is the new human condition better or worse than the old one?

Is the new human condition an improvement on the old one or a step backward? It is hard to say until we live it. It will almost certainly require new ways of coping with time and mortality. The essence of the new human condition is that death is unscheduled and therefore not a hard limit.

In the existing human condition, death *is* a hard limit; we never get much more than a century of life and must arrange our lives within that schedule. We have a sense of how much sand is left in our hourglass, and we tend to see our lives mapped out against a timeline, almost as if watching ourselves moving along that timeline from birth to death, the future growing shorter and the past growing longer. In this human condition, life is a bit like a movie; we have a sense of the beginning, middle, and end, and even if the plot is a confused and chaotic mess, we know roughly when it will be over. If you know you're likely to live as long as your 80s or a bit beyond, you can plan a career where you rise to seniority in your field, achieve what you believe you have the capacity to achieve, watch your children produce grandchildren, and watch your grandchildren grow up. We all know this movie, and many of us hope to star in it.

Some writers suggest that life has meaning when it has the kind of structure a fictional narrative has, where the parts of the story cohere into a single overarching theme and gain their significance from their relationship to that theme. For example, I started my professional life as an attorney, hated practicing law, and after 17 years of this, I managed to change professions and become a philosophy professor. I used to think my years in law practice were wasted years, but I now see some value in them. Learning to function competently in a profession that didn't suit my temperament or values made me a broader person, gave me some skills I would never have acquired otherwise, bolstered my self-confidence, and made me appreciate the fact that we're all capable of a far wider range of activities than we think we are. Those years now serve as an episode in a life story that (like many life stories) involves struggle to overcome setbacks, along with increased self-awareness and self-realization. My years in law practice gain significance and meaning from the place those years have in that larger narrative. (That's not to say I would do it all again.)

So the meaning of a life often comes partly from its narrative structure. However, just because it can be meaningful in that way does not mean that it can be meaningful only in that way. We are used to such narrative structures because they fit the life span we are used to. In an extended life bounded only by unscheduled death, we might develop other structures or find that large units of time can be narratives in their own right. A very long life might contain many movies, as it were, or resemble a miniseries with no preconceived conclusion. Or we might find meaning not in the unfolding of a finite story but in something entirely different, such as deeper spiritual development. The finite story kind of meaning works for lives whose meaning is tied up with career and raising a family, but we might move past those concerns and find our meaning elsewhere, somewhere outside ourselves.

In the new human condition, we might try a series of careers or different ways of life with no intention of assembling it into a coherent story. Each of those phases might have meaning or value without being arranged into any larger pattern. Moreover, the new human condition may (paradoxically) require us to live more in the moment, despite having more time. Living without a timeline and without any sense of how much time is left may force us to value the present and not worry too much about the future, partly because the odds of death are always low (compared to what they are now in old age) and partly because we live long enough to accomplish a lot and therefore don't fear that we'll die before we experience life and achieve some major goals. It may be easier to accept death if we get enough time to correct the mistakes of our youth and accomplish what is important to us.

The new human condition will be very different from the old one, but once we get used to it, it might be better than the old one—and not just because it lasts longer.

3.11 Conclusions

Here are my main conclusions about the issues raised in this chapter, continuing the letter order from the conclusions listed at the end of chapter 2:

D. There are reasons to believe that extended life would be just as good as life in a normal life span, all things considered. (Section 3.2)

E. The bioconservative arguments tend to exhibit four mistakes: they use arguments that work only (if at all) for immortal life, they overlook counterintuitive implications that arise when we extend those argument to all forms of immortal life, they cannot be run backward without absurd results, and they look only at the possible downside to extended life. (Section 3.3)

F. There is insufficient reason to believe that extending the human life span will make it harder to accept death, undermine our motivation and make us waste time, be less meaningful than life as we know it, or make it harder to develop virtues and care about things beyond our selves. (Sections 3.4 through 3.8)

G. Extended life presents a new human condition, especially in radically extended lives. The new human condition has four features: (1) aging is elective, (2) life extension is reversible, (3) death will be unscheduled, (4) your life expectancy at any given age will always the be same, and death will deprive you of vastly more time no matter how old you are when you die. The new human condition might be better than the old one. (Section 3.10)

4 The Will-nots—Life Extension and Suicide

4.1 Introduction

Many people say they aren't interested in having an extended life. They are potential Will-nots. (They would be actual Will-nots if life extension existed and they could afford it but turned it down anyway.) For various reasons, they think extended life is not a good life. I discussed some of those reasons in the last two chapters. I now want to discuss another reason they might have for thinking that life extension is bad for them. I am not talking about Malthusian pressures or other consequences that are bad for everyone, including the Will-nots; I'll get to those issues in the next two chapters. I am talking about possible harms that affect only Will-nots. Moreover, these harms—if they exist—affect Will-nots even if they never use life extension. In short, I will discuss possible reasons for thinking that merely *living in a world where life extension is available* is bad for Will-nots, even though they never extend their own lives. That kind of harm is important, for we can't answer it by saying, "Fine, maybe extended life is not good for you, but that's no reason to stop anyone else from getting it." If living in a world where life extension exists is bad for Will-nots even though their lives are not extended, then we have some reasons against developing life extension even though it's optional and reversible.

Do such reasons exist? This concern has an obscure provenance. I confess that I can't offer any quotes or citations for it, and I've never seen it explicitly mentioned in any publication (though I feel I sense it somewhere between the lines in certain bioconservative passages). However, when I talk to people who don't want life extension, I sometimes have the sense that some of them don't merely want to avoid living an extended life. They seem to feel distaste or apprehension at the very idea of living in a world where life extension exists at all. There is something about living in such a world that feels uncomfortable to them—and not just because they're worried about overpopulation or social injustice or because such radical change may seem threatening. In this chapter, I'll try to articulate and explain two reasons why they might legitimately feel that way.

First, they might think that the mere fact that life extension is available reduces the advantages of living a normal life span, at least for them. To put this another way, it might reduce the death benefits of their normal life spans. (I didn't argue in chapter 3 that there are no such benefits, only that they're not as numerous or significant as some thinkers believe.) Second, they may think that refusing life extension or discontinuing it after they started it is a kind of immoral suicide so that creating life extension puts them in the position of having to choose between living an extended life they don't want and doing something immoral.[1]

By the way, the suicide question is not an issue only for Will-nots. It is also an issue when I claim that we should not worry about whether extended life is good to have, for we can always go off our life extension meds and resume aging (in effect, transferring from the Haves to the Will-nots). If doing that is an immoral suicide, then my arguments in favor of developing life extension are undermined in a more general way.

I will argue that neither of these reasons is very strong.

4.2 Would making life extension available reduce the death benefits for Will-nots?

As we saw in the last two chapters, some people believe that aging and a normal life span have advantages. These advantages, which I call "death benefits," might include greater ease in accepting death, using our time better, having a certain kind of meaning in life, developing virtues, learning to transcend our own narrow concerns by developing concern for posterity and other things beyond ourselves, and avoiding the risk of intolerable boredom. (The concern that you may eventually change enough to become a new person, and thereby cease to exist in the long run even if your body survives, is not an objection to extended life so much as a reason to think you cannot extend your life forever.) These are offered as reasons not to extend your life: extend it, and you'll lose these benefits. Now, imagine that you can afford life extension, but you turn it down to avoid reducing those benefits. You are a Will-not. You expect to die in your 80s or 90s, and you want to enjoy your death benefits in their full measure. If life extension is available to you, then you might feel that your death benefits have been reduced somehow even if you refuse it, choosing to age and die on schedule.

Here is why you might feel that way. Without life extension, death is a hard limit—you can avoid it for a while, but you will run up against it in roughly a century or less. However, as I argued in section 3.10, if life extension is available to you, then for you, death is no longer a hard limit. It is a soft limit in the sense that it's a limit you can postpone if you choose to take advantage of life extension. This is true not only if you extend your life but also if you don't, for if you refuse life extension when it's available to you, you can later change your mind and extend your life, thereby postponing death after all. You have this choice during much of your life, or at least until you're too elderly to make use of life extension (and perhaps even then, if we learn to reverse aging). The significance of this is that the death benefits depend partly on death being a hard limit. Thus, the fact that you can change your mind and start extending your life—even if you never do—might seem to reduce some of your death benefits.

Does it? Of course, that depends partly on whether there *are* any death benefits. The death benefits are the flip side of drawbacks to extended life. Having a normal life span is beneficial because you avoid the drawbacks. I argued that these drawbacks are either minor, nonexistent, or avoidable—or, at very least, we lack evidence that they are serious and unavoidable. I also noted that the advantages and drawbacks of extended life may vary from person to person. However, in this section I'll assume, just for discussion, that extended life has some drawbacks for at least some people. With that in mind, let's consider how living in a world where life extension is available might reduce some death benefits for some Will-nots.

First, you may not come to terms with the inevitability of death if death is not a hard limit. If you can always change your mind and start using life extension, then death may not feel inevitable, and it may be more difficult to accept death. Thus, living in a world where life extension is available to you may undermine the first death benefit even if you never use life extension at all. Second, if you can get life extension when you want it, then you might procrastinate and use your time poorly even if you never extend your life. After all, you can always take the cure and buy more years if you feel the press of time. Just having that option might induce you to waste time. Third, if the virtues we develop by facing mortal danger, such as courage, are harder to acquire in an extended life, they may also be harder to acquire if you think you could still delay death substantially if you changed your mind and used life extension. Presumably the same is true of death's power to drive us to transcend our narrow, selfish concerns and invest ourselves in posterity and other things that are bigger than we are. As for avoiding the boredom of extended life, that's one drawback the Will-nots will completely avoid, provided they never change their mind and join the Haves.

So the advent of life extension might reduce some death benefits for some Will-nots. (The advent of life extension would not, of course, have this effect on Have-nots who don't want life extension, for the fact that they can't afford life extension makes it effectively impossible for them. For them, death is still a hard limit.) How serious a harm could this be? We have no data and no algorithm for calculating that harm, so all anyone can do is use their judgment and hazard a guess. My guess is that it's trivial, but perhaps the magnitude of this harm varies from person to person.

4.3 If you refuse or discontinue life extension, are you committing suicide?

In any case, there is a second way in which the advent of life extension might affect the Will-nots: it might force them to choose between (a) an extended life they don't want and (b) committing what they consider an immoral kind of suicide. Does introducing life extension into the world put such people in a position where they are damned with extended life if they take the cure and damned for committing suicide if they don't?

First we must ask whether declining or discontinuing life extension *is* a kind of suicide. We can approach that question by drawing an analogy between forgoing life *extension* and forgoing life *support*, such as respirators, feeding tubes, or intensive care. No one has foregone life extension yet, but people forgo life support in hospitals all the time. There are two further questions here:

1. Is foregoing life support a form of suicide?
2. Is foregoing life extension morally equivalent to foregoing life support?

If the answer to both questions is yes, then foregoing life extension is a kind of suicide. If the answer to either question is no, then foregoing life extension is not a kind of suicide. (My discussion treats never starting life support the same as stopping life support after it has started.[2]) I ask the reader to be patient while we take a short detour and consider life support.

Is forgoing life support a kind of suicide?

Suicide consists of intentionally bringing about your own death. The US Centers for Disease Control and Prevention define suicide as "death caused by self-directed injurious behavior with any intent to die as a result of the behavior,"[3] and the Oxford Dictionary of American English defines it as "the action of killing oneself intentionally," noting that the word comes from the Latin words *sui* ("of oneself") and *caedere* ("kill").[4] These definitions are a bit terse; here is a similar but better one from Michael Cholbi:

Suicide is intentional self-killing: a person's act is suicidal if and only if the person believed that the act, or some causal consequence of that act, would make her death likely and she engaged in the behavior to intentionally bring about her death.[5]

Cholbi says "intentional self-killing" occurs only when "the individual rationally endorses her own death as the chosen means or a foreseeable effect of pursuing her ends."[6] The agent need not endorse death as such; it is enough that she endorses dying when that's the only way to achieve some end she values more than staying alive.

Note that all three definitions are purely descriptive. They leave it open whether suicide is always immoral, never immoral, or sometimes immoral—and when. This is as it should be; we need a concept for intentionally bringing about your own death, and we should not assume that doing so is always wrong. Even if it *is* always wrong, this does not follow from the concept of suicide but must be argued from other premises.

Given our concept of suicide, it turns out that declining or terminating life support can count as suicide. Suppose, for example, that you're dying slowly from a very painful, incurable, and untreatable cancer and that your life support includes a ventilator. You act by telling your doctor to turn off your ventilator, believing that this will make your death likely, and you do this in order to bring about your death and thereby avoid unbearable pain and what you consider a meaningless life. This fits all the elements of the definition: you acted in a way that made your death more likely because you wanted to bring about your death, which you rationally endorsed as a means of avoiding pain and a meaningless existence. The same analysis should work for many ways of declining or discontinuing some treatment that would keep you alive, such as declining cancer treatment knowing this will shorten your life by six months, provided you did so in order to bring about your death and believed that doing so would make your death more likely. (This is why merely neglecting your health to the degree that you foreseeably shorten your life is usually not suicide: people who neglect their health typically don't do so in order to hasten their death.)

Cholbi has argued this way.[7] So have Franklin G. Miller, Robert G. Truog, and Dan W. Brock, who define suicide as aiming at and causing one's own death, and contend that this happens when, for example, patients request to be taken off a ventilator to avoid life as a quadriplegic.[8] David Shaw argues that the body is a kind of life support for the person (or at least the brain), and thus euthanasia is a way of taking someone off life support—and hence morally permissible. Shaw does not discuss suicide, but it's easy to extend his argument to suicide. Euthanasia involves killing someone else, while suicide involves killing yourself.[9]

One might resist the claim that foregoing life support is suicide by appealing to the killing/letting-die distinction: suicide involves killing, while foregoing life support is a form of letting-die; therefore, forgoing life support is not suicide. This is not because death happens more slowly when you let yourself die and more quickly when you kill yourself; you can kill yourself by ingesting a poison that takes months to kill you. Rather, the objection is that suicide requires killing and not merely letting-die.

One possible argument behind this objection is that suicide is necessarily immoral, and letting-die is not immoral, so letting-die is not suicide. However, that can't be right. Aside from doubts about whether the killing/letting-die distinction is morally relevant at all, it should be uncontroversial that some instances of letting-die are clearly immoral too, such as removing your competent grandmother from her respirator in order to speed up your inheritance. It should also be uncontroversial that some instances of killing are not immoral, such as killing in self-defense.

Another possible argument for this objection appeals to how we use the concept of suicide. Here the claim is that we use the concept of suicide only in cases of killing one's self, and never in cases of letting one's self die, so that killing is simply part of the concept of suicide, whether or not the killing/letting-die concept makes any moral difference here. Now, it is true that we rarely use the term *suicide* to refer to cases of letting-die. However, there are two possible reasons why we do this, and neither one is a good reason to say that letting-die can never be suicide.

One possible reason is that in most cases of letting-die, the decision is made by a surrogate decision-maker and not by the patient himself, who is usually incompetent in such cases; this is not suicide because suicide requires action by the one who dies. However, that doesn't preclude applying the concept of suicide to cases of letting-die where the decision to let the patient die is made by the patient. Another possible reason is that there are laws prohibiting helping someone commit suicide, yet in most cases of letting-die, the medical staff help, at least by terminating the treatment that sustains life. Medical professionals don't want to call this suicide, partly to avoid upsetting the patient or family and partly to avoid attracting attention from prosecutors. However, these are pragmatic concerns and not good reasons to limit the concept of suicide to cases of killing.

Moreover, not all cases of bringing about death fall clearly on either side of the killing/letting-die distinction. Consider some cases.

First case: I have accidentally ingested poison with no intent to die. Now I have only a few hours to take the antidote, and I decide to forego the antidote because, on reflection, I've decided to die in order to get a big life insurance payment for my family.

Am I letting myself die, or am I killing myself? Does it matter that I refrain from ingesting the antidote?

Second case: If I stop taking some other medication I've been taking every day, the poison will do nothing—it works only in conjunction with that other medication. I then take the other medication in order to make the poison active, and thereby die.

If that's suicide, then why isn't refraining from taking an antidote just as much a case of suicide?

Third case: I'm on a respirator. I reach over and pull the plug out of the socket, turning off the respirator.

Taking someone off a respirator is usually considered a case of letting-die, but if I did this to someone else, would I not be killing him? If so, then I'm killing myself when I do it to myself.

The point of these three cases is that there are clear cases of suicide that do not fit easily on either side of the killing/letting-die distinction. It therefore makes sense to characterize suicide more neutrally as "bringing about death." Therefore, both forgoing and terminating life support can be suicide.

The detour is over. Back to life extension.

Is forgoing or discontinuing life extension morally equivalent to foregoing life support?

Yes: they are morally equivalent. If it is suicide to decline or terminate chemotherapy and die from cancer, thereby giving up six months of life, then it is even more clearly suicidal to decline more radical kinds of life extension that may confer centuries of life. The fact that the potential life is longer and better does not make this less a case of suicide; if anything, the reverse. I'm not saying that life extension is a kind of life support, only that there's no morally relevant distinction, so far as suicide is concerned, between life support and life extension—if terminating one counts as suicide, so does the other. Both of them require help from medical professionals, both are beyond the natural capacity of the body to heal itself, both will make your life longer, both can be declined, and both can be discontinued.

Here is an objection: the definition of suicide refers to bringing about one's death, but "death" here refers to death at the end of our natural, biological life span, not to death after an unnaturally long, artificially extended life span. In other words, "bringing about your own death" means "bringing about your own death before the natural end of your life," not "bringing about your own death when it is possible to live longer than we naturally do." Once you reach, say, age 100, you have a free pass to check out any time you like without being labeled a suicide for doing so.

What could be the argument for this claim? I suspect that those who find it appealing are drawing upon a concept of suicide as something bad and further drawing upon a sense that what is contrary to nature is bad. If aging and dying in a more natural way is right and suicide is wrong, then turning down life extension must not be suicide. Once again, this kind of argument relies on a normative definition of suicide as something necessarily wrong or bad, and for the reasons given earlier, we should use a purely descriptive definition of suicide rather than building a normative view of suicide into the definition. Claims that suicide is wrong must be argued for, not assumed in the definition. Moreover, the claim that what is natural is presumptively right and what is unnatural is presumptively wrong is subject to too many counterexamples to have much weight as a moral rule of thumb: vaccination, eyeglasses, and literacy, among many others. (We'll discuss the significance of what is natural at greater length in sections 12.5 and 12.6.)

4.4 If refusing or discontinuing life extension is suicide, is it immoral?

I have argued that declining or terminating life extension is a kind of suicide. Now let's consider whether this kind of suicide is immoral. Of course, this question is too broad; except for absolutists who believe all suicides are always wrong, the morality of suicide depends on various factors. Let's ask a narrower question instead: Is the fact that one commits suicide by refusing life extension *relevant* to whether or not that suicide is morally permissible? In other words, does the fact that life extension is involved ever make a difference to the morality of suicide, or is it, like the difference between using a gun and using poison, morally irrelevant? I think it's morally irrelevant. This is not to say that suicide by refusing life extension is always morally permissible, and certainly not that it is never permissible—only that the morality of suicide by turning down or discontinuing life extension turns on arguments about the morality of suicide in general.

To see why, consider the main arguments against suicide:[10]

• It violates natural law. Aquinas argued that suicide is contrary to natural law because all living things are naturally inclined to try to survive.
• You are God's property. If you kill yourself, you are destroying God's property.
• Life is a gift from God, and committing suicide is ungrateful.
• Human life is intrinsically valuable, and you should never destroy something with such great intrinsic value.
• Suicide denies society the benefits of your continued existence. Society will be worse off if you take yourself out of society by killing yourself.
• You have a debt to society because society made your life possible and conferred various benefits on you. You have an obligation to stick around and repay society.

• You have responsibilities to others. Suicide prevents you from fulfilling your obligations to your children, your other family members, and anyone else you have a responsibility to care for.

• It is using yourself as a mere means; Kant argues that committing suicide uses a person (you) as a mere means to something else—your happiness.

None of these characterizations do justice to these arguments, but they provide enough detail to see whether and how life extension might be relevant to arguments that suicide is immoral.

The fact of discontinuing or refusing life extension seems to make no difference to the natural law argument. Life extension is artificial and therefore unnatural, in a sense, but that artificiality seems irrelevant to whether there is a natural law mandating survival. Indeed, the survival instinct is one of the main impulses behind life extension research, though few researchers will say so openly. If it's immoral to refuse life support for this reason, then it's also immoral to refuse life extension. However, it's neither more nor less immoral simply because life extension is involved. The fact that this occurs by turning down life extension is morally irrelevant.

Turning down life extension also makes no difference to arguments that you are God's property or owe him a debt of gratitude. If being his property or in his debt makes suicide wrong, it makes suicide wrong however you commit it. The fact that life extension is involved is morally irrelevant here.

As for the argument that human life has intrinsic value, if a natural life span has sufficient intrinsic value to make suicide wrong, then an extended life span will too. Some may say that extended life is less valuable to the person living it the longer it continues. This is a version of the boredom, personal identity, and death benefits arguments. However, that's a claim about the value of your life for you, not a claim about the intrinsic value of your life. Again, even if suicide is wrong for this reason, the fact that life extension is involved does not make the suicide more or less wrong.

Now consider the arguments based on your relationships and obligations to society and others. The fact that life extension is involved is irrelevant to those arguments too. Your relationships, obligations, and value to society do not go away just because you live a lot longer, so if those things make suicide immoral in a normal life span, they should do so in an extended life span too. The same seems true of using yourself as a mere means, for halting aging doesn't mean you are no longer an autonomous person.

In general, the fact that you're turning down life extension seems irrelevant to whether suicide is wrong. If any of these arguments establishes that at least some suicides are morally wrong, then at least some cases of declining life extension are morally wrong too.

The fact that life extension is involved might, however, be relevant to the degree of wrongness (if any). This depends on the argument. Many of these arguments make some reference to the value of your life to others (society, your children, God) or its intrinsic value. If suicide is wrong for one of these reasons, then it may be more wrong the more potential years of life you are avoiding, for life extension makes so much more of that valuable life span available. However, the fact of life extension probably makes no difference to the degree of wrongness when it comes to arguments that suicide is wrong because it violates natural law or that you are God's property. If it violates natural law, that violation does not seem more serious the longer the span of time one fails to survive. The fact of life extension is also probably irrelevant for the Kantian argument that committing suicide is a case of using yourself as a mere means, for it is not clear how the number of years involved is relevant, unless the Kantian wants to say that you are using yourself as a mere means for a longer period of time the longer your potential life span was.

So the fact that it's irrelevant that suicide occurs by refusing life extension does not tell us whether refusing life extension is morally wrong, but it might be wrong for some other reason. As I said, whether or not suicide is morally permissible depends on the case, so it's hard to generalize, but we can try to identify what might make it wrong and the odds that it would ever be wrong. If the odds are low, then we have one fewer reason to think that some Will-nots may suffer harm by being forced to choose between extended life and what they regard as suicide.

To evaluate this, I want to divide the arguments against suicide into two groups. Some of the arguments appear to imply that suicide is always wrong. These arguments—that it violates natural law, or that we are God's property and life is a gift from God, or that human life is intrinsically valuable, or that we should not use ourselves as a mere means—don't seem to be limited to particular cases. When people say life came from God, they mean we all got it that way, not just those who are on particularly good terms with God. When people say human life has intrinsic moral value, they typically mean that all human life has this. When we're taught that it's wrong to use other people, we are not told that it's wrong under certain circumstances but perfectly all right under other circumstances. We are told that it's wrong, period. These arguments appear to be *absolutist*; they seem to say that suicide is always wrong. I want to set aside the question whether those views are correct and ask what they imply about refusing life extension. They might seem to imply that refusing life extension is always wrong.

Then again, they might not imply this. It may be that suicide is always wrong but that sometimes there are reasons for doing something that happens to be suicide that are so strong, they outweigh the wrongness of suicide. We think this way in many situations. For example, many of us would say that it's not wrong to lie to a murderer about the location of his potential victim. It's still lying, of course, but even if lying is always wrong, telling a lie might, for countervailing reasons that are more important—like preventing a murder—be the right thing to do. We might say something similar about refusing or discontinuing life extension. For example, we might say that when you have strong reasons to prefer not to continue living, refusing or discontinuing life extension is still suicide, but the moral reasons against doing so are outweighed by other moral considerations. For example, if you find that the drawbacks of extended life are pretty bad and you're bored beyond toleration and longing for death, then you might have moral justification for committing suicide by refusing life extension despite the fact that doing so is wrong.

I said that the arguments against suicide divide into two groups and that the first group is a set of arguments that, by their terms, seem to apply to all cases of suicide. The second group of arguments do not, by their terms, apply to all cases of suicide. They apply to certain cases and not others. One of these arguments says that suicide denies society the benefits of your continued existence and that society will be worse off if you take yourself out of society by killing yourself. This argument is conditioned on the idea that society will suffer harm if you disappear, but not everyone is that essential to those around them, so this argument does not seem to apply to all cases of suicide. Another argument says that you have a debt to society, for society made your life possible and conferred various benefits on you. You therefore have an obligation to stick around and repay society. This too depends on circumstances. The argument probably doesn't apply in cases where you long since repaid the value of whatever society gave you. Similarly, arguments against suicide on the grounds that you have responsibilities to children, other family members, or friends are only as strong as those responsibilities. If your friends and family are already well situated and no longer need your help, such arguments lose force.

You can see where this is going. Suppose you're elderly and the available life extension methods can actually reverse aging, so it's still not too late for you. Nonetheless, you refuse life extension. This could be wrong if you still have responsibilities to people around you or debts to repay. I'm not saying that elderly people never have such responsibilities or debts. I do claim, however, that by the time you're elderly, you may well have done all you're required to do for your children, relatives, or society at large. At some point, those responsibilities come to an end.

So is it morally permissible to refuse or discontinue life extension in most cases? I am inclined to say yes, if only because a normal life span is usually long enough to fulfill responsibilities and pay debts. (If it were not, we would also think that refusing life support is always wrong, but we tend not to think that.)

However, even if I'm right that refusing life extension is rarely an immoral suicide, Will-nots who think it's an immoral suicide may suffer a harm anyway. I'm talking about "moral injury," defined as "perpetrating, failing to prevent, bearing witness to, or learning about acts that transgress deeply held moral beliefs and expectations."[11] The concept of moral injury was developed by psychiatrists working with war veterans who had participated in actions that violated their moral ideals. Victims of moral injury suffer guilt, shame, self-condemnation, demoralization, and self-harming behaviors. The injury is that you suffered such feelings because you were led to behave in ways that conflict with your moral beliefs.[12]

Some Will-nots may believe that refusing or discontinuing life support is a kind of immoral suicide. To suffer a moral injury, it's not necessary that your moral belief is true, only that you sincerely believe it and you are led to act contrary to that belief. The harm lies in the guilt, shame, and so forth, not in the supposed wrongness of what you are doing. Therefore, Will-nots can suffer a moral injury even if they're mistaken about whether declining life extension is suicide or about whether such suicide is immoral. Because introducing life extension into the world and making it available forces that choice on the Will-nots, developing life extension imposes a kind of moral injury on those Will-nots.

Let me add four qualifications. First, not all Will-nots suffer this moral injury—just the ones who believe that turning down life extension is an immoral suicide but turn it down anyway. The percentage of Will-nots in this group may turn out to be fairly small.

Second, I have defined *Will-not* as someone who has access to life extension but turns it down. Of course, there may well be many Have-nots who don't want life extension and would turn it down if they had access to it. Those Have-nots are similar to Will-nots, but I leave them out of the Will-not category because they have no choice when it comes to using life extension (it's not available to them), so they're not forced to make a choice that's contrary to their moral beliefs.

Third, it's possible that some people who have access to life extension don't want it but accept it anyway, for they believe that doing otherwise is an immoral kind of suicide. I doubt that very many people will fall into this category, so I have not set aside a separate subcategory for them. However, anyone who falls into this category suffers a kind of injury too, for although they haven't committed suicide by turning down life extension, they are living a life they don't want and presumably don't like. That may not be a moral injury, but it is an injury.

Fourth, I have repeatedly pointed out that life extension is a reversible decision, so if it turns out to be boring or otherwise undesirable, a Have can always reverse that decision by going off her life extension meds, as it were, and resume aging until she dies in a few years or decades. My argument that this is not an immoral suicide is therefore an important element in my moral arguments for life extension in general, for the option of discontinuing extended life is a sort of escape clause for those who find they do not like life extension. However, Haves who do that are effectively becoming Will-nots, so if they believe that discontinuing life extension is an immoral suicide, then what I've said about the morality of committing suicide by discontinuing life extension applies to them too.

4.5 Conclusions

H. To the extent there are benefits to living a normal life span, refusing life extension but living in a world where life extension is available might reduce death benefits for the Will-nots to some degree. However, any harm of this kind is likely to be very minor. (Section 4.2)

I. Refusing or discontinuing life extension is a kind of suicide, but not an immoral suicide. (Sections 4.3 and 4.4)

J. Will-nots who live in a world where life extension is available and who believe that refusing or discontinuing life extension is an immoral suicide may feel that they've been forced to choose between an extended life they don't want and a form of suicide that is morally wrong. Will-nots who believe this suffer a harm called "moral injury" (though not all Will-nots have that belief about life extension and suicide, so not all of them suffer a moral injury). (Section 4.4)

5 Everyone—Social Consequences

5.1 Introduction

What would it be like to live in an extended world, a world where lots of people live extended lives? I'm not asking what it's like to live an extended life; we discussed that in chapters 2 and 3. I am asking what the *world* would be like, for everyone, if a large portion of the human race used life extension. Of course, the answer depends partly on how many people have access to life extension. If very few people can get it, the consequences will be less dramatic. If lots of people can get it, the consequences will be more dramatic. Let's imagine an extreme case and assume that anyone can get it and a large portion of the human race decides to use it. How would that change the world, not just for particular groups such as the Haves, Have-nots, or Will-nots but for everyone? (In chapters 7 and 8, I'll discuss issues that are specific to the Have-nots.)

In section 5.2, I'll survey several potential bad social consequences that opponents of life extension often mention. They include, among other things, possible problems with pensions and supporting retired people, changes in the relations between older and younger family members, the possibility of dictators living forever, and what I call "opportunity drought," where older people hang on to property and position forever and ever. I'll argue that these are not insurmountable problems, and they're not strong reasons against developing life extension. In section 5.3, I'll try to balance the conversation by pointing out several possible good consequences from widespread life extension. These include a human race with enough time to become wiser, better-informed, more concerned about posterity, and less violent. They also include the possibility of developing new opportunities and new and desirable ways of living we can't foresee now.

In this chapter, I won't discuss the most important problematic consequence widespread life extension might cause: a Malthusian crisis. If people live far longer, the earth may become overpopulated, partly because
people fail to die on schedule and hang around a lot longer and partly
because they have more time to have more children. This could put catastrophic pressure on the world's resources and ecosystem. This consequence is so important, and the issues it raises so complex, that it needs
a chapter of its own. I take this up in chapter 6. (In chapter 8, I take up
the other major problem: life extension may be so expensive that only the
rich, or the middle class, can afford it.)

As for the consequences I do discuss in this chapter—good, bad, or
indifferent—bear in mind that radical changes are coming whether or not
life extension becomes widely available. Automation, robotics, artificial
intelligence, nanotechnology, genetic engineering, virtual reality, and self-
driving cars are just a few of the technologies that are already turning the
world upside down, and most of these technologies promise both benefits
and challenges. The human race has a long history of adapting to radical
changes of all kinds—from the advent of domesticated animals through the
development of agriculture, writing, industry, and technology—ultimately
leading to a world of colossal cities in a "global village." Arguably all of
those developments have both an upside and a downside. At very least,
they all present challenges. Life extension will be no exception.

5.2 Potential bad social consequences

Many people, when they think about life extension for the first time, see an
array of possible social problems.[1] Most of these problems are not all that
difficult to handle; the human race has surmounted far greater challenges.
As we sort through them, ask yourself this: Is living in a world with those
problems a life worse than death? Would you prefer to die earlier in order
to avoid living in that world? That is, after all, the alternative.

Relations among family members and generations
In the movie *In Time*, everyone halts aging at around 25, so the hero (played
by Justin Timberlake) has a mother (played by Olivia Wilde) who looks as
young as he does (they make a weirdly attractive couple). We can take this
a bit further: Imagine trying to keep track of not just 3 or 4 generations of

relatives, but 10 or more. Imagine meeting your great-great-grandmother, and she looks like your contemporaries. Imagine trying to maintain anything like a traditional relationship with your children when you and they enjoy the health and appearance of a young adult and the generations pile up, one after another, with no one obviously qualified for role of an elderly counselor, for no one looks older and everyone has time enough to gain the wisdom of age. It would take some time to adjust to these situations and work out new ways of interacting with each other, but doing so does not strike me as a life worse than death. These adjustments don't even seem particularly difficult once you consider that we have already gotten adjusted to same-sex marriage, gender transition, surrogate mothers, effective birth control, and other novelties.

Of course, lifelong marriage may not survive all this. It's one thing to vow to stay together until death when you expect death in 50 years or so. It's quite another to make that commitment when death may be centuries away. However, there are other possible forms of marriage: marriages intended to last a fixed period of time, with an occasion for recommitment every so often, or polyamorous relationships that involve a shifting collection of individuals, to take two examples. Humans have had many different forms of marital relationship, and we may find that we're more limber than we realize. You could, of course, as part of your marriage vows, vow that neither of you will use life extension and that you will grow old together on a traditional schedule: 80 years or so. There's nothing to stop you from doing this if it means that much to you.

There's not much more we can say about this: your imagination is as informative as mine. My take on this is that extended life is just another massive change in the human condition, on a par with what we navigated as we moved from being hunter-gatherers to living on farms or from farms to the industrial and technological world of the city. We'll work it out, bit by bit, over time.

Entrenched elders and opportunity drought

Many people worry that dictators will extend their lives; we'll be stuck with North Korea's current ruler indefinitely, along with Syria's president Assad and any number of lesser tyrants, including workplace bosses and anyone else with enough power to make life miserable for those under them. Tenured professors may keep their jobs and crowd out younger thinkers. Land may rise in price as older generations hang on to it. Promotions may be harder to get. Middle and upper management may become thick with people who have generations of experience, and workplaces might practice age-discrimination against the young—or the younger.

But then again, maybe not. Tyrants can be overthrown or killed, and we need not wait until they're elderly. (The vast majority of Roman emperors died in office—and not from natural causes.) Tenure can be eliminated, if that becomes a problem. New business enterprises can be founded to compete with old ones. Microsoft was not taken down by an antitrust lawsuit, but it *has* been undermined by new companies exploiting new technology, like Google. Laws can be passed to institute land reform, forcing it out of the hands of those who've held it long enough (a kind of geriatric eminent domain), much the way land reform transfers land from elites to peasants. We already set a time limit on patents; we can do the same for ownership of other things.

As for the career advantage of having lived a lot longer than other people, this may be overrated. I have some personal exposure to three professions: teaching philosophy in a college, practicing law, and medicine (I never practiced medicine, but my first academic job was teaching medical ethics in a medical school). For whatever it's worth, my impression was that in most professions—even very demanding ones—most people get about as good as they ever will in the first 20 years. Additional experience adds something, but less than you might think; the law of diminishing returns applies here.

The rate of innovation and social change

Some writers have warned that a world full of people who live extended lives will become increasingly conservative and resistant to change. It's true that young people tend to be the agents of social change and more open to novel approaches and ideas. Creative and innovative industries tend to be led by people in their 20s and 30s. Of course, we might be pleasantly surprised. We might find that there is still enough innovation and social change going on. Even now, some people remain actively committed to social change or creative work all the way into their 80s and might do so even longer if they had more time. In fact, we may find that some people enter completely new phases of life when they live long enough; who knows what personal growth we may achieve when we live for centuries?

Still, the critics may have a point. Most of us form our worldview by our mid-40s or sooner, and it doesn't seem to change much after that. Still, what price would we pay to preserve this? Is living in a world that has grown more conservative in this way a life worse than death? Run the argument backward: Suppose we could achieve an even more innovative, fast-changing, creative world by taking medication that produced accelerated aging after age 35—two years of senescence followed by death at 38. This would get rid of all the older generations who slow everything down. If 35 seems too young as a cutoff age for avoiding the conservatism of older generations, then suppose we could devise a drug that kicks in as soon as you stop forming new neural pathways so that you start aging as soon as you become less open to new ways of thinking. Those who keep their creative edge longer get to live longer. Would you vote for that policy? Now run the argument in the other direction: If you would not shorten your life to get that result, why would you refuse to extend it to get that result?

It's also possible to overestimate the value of innovation and change. During most of human history, the pace of change was glacially slow. Most people living any time before the last three hundred years or so would not have noticed much change during their lifetimes, aside from new cycles of war, famine, and conquest. Moreover, although rapid innovation is exciting for some people, many people find it very stressful. There's a lot to be said for tranquility and predictability.

Pensions and elder care

Will we end up with huge numbers of elderly people needing a lot of expensive care from a shrinking pool of younger people? One fear is that people will retire at 65 and live to 165, straining the pension system to the breaking point.

The first thing to note is that we may already be facing that problem, even without life extension. The world's population is becoming older on average, and in many countries, there are fewer young people to support a growing number of retirees. Fertility is falling below replacement levels almost everywhere (except Africa, for another century), and the population in many countries is shrinking and becoming more and more gray. In 2015, 12 percent of the world's population were over 60 (24 percent in Europe); by 2050, one-quarter in all regions outside Africa will be 60 or over.[2]

The older the average age, the harder it is for a society to support its elderly. The number of workers per retiree is known as the potential support ratio (PSR). According to one team of demographers, America's PSR will drop from 4.6 today to 1.9 by 2100.[3] Here are comparable figures for some other countries:

	Workers per retiree today	Workers per retiree in 2100
Brazil	8.6	1.5
China	7.8	1.8
Germany	2.9	1.4
India	10.9	2.3

Note that all these countries (and many others) will have fewer workers per retiree than Japan currently has (2.6 workers per retiree today), and Japan is already noteworthy for having so few that supporting its retirees is a serious challenge.

This trend poses a severe challenge for pensions and other institutions that provide for the elderly: there just aren't enough young people to easily care for the old people. How would life extension affect this? It probably won't make this any worse. Suppose, for example, that we slow aging to half its current rate and that the elderly portion of a life span is the same portion as now (roughly 15–20 percent). In that case, instead of needing a pension and care from age 65 to 85 in an 85-year life span, we would need pension and care from age 130 to 170 in a 170-year life span, but there would be as many younger people supporting that group as there are now.

Life extension might even help solve this problem. Suppose, for example, that life extension compresses morbidity—in other words, that the percentage of life we spend in an elderly condition is smaller than it is now. For example, in that 170-year life span we might be in an elderly condition for only the 20 years or so that we are now—or even fewer, perhaps 5 or 10 years. In that case, the number of healthy working people supporting each individual too old to work will be larger than it is now, and the pension crisis will be easier to handle.

Finally, if we halt aging altogether, we will not become elderly at all. In that event, the only people who need to be cared for are those who suffer a serious physical or mental disability, not because they are old, but for other reasons. Of course, you might not want to work forever and ever and ever, but the fact that you do not age does not stop you from taking a break. You will have abundant time to save money, pay off your mortgage, get all your kids through college, and take a long sabbatical from the working world if you wish, coming back into the workforce when your savings are depleted.

Here's another pension-related objection to life extension that comes up now and then: Even if life extension halts aging in a youthful phase of life, what about all those lifetime pensions people are entitled to under social security and various private pension funds? We can't keep paying social security benefits forever to someone who worked 40 years and then lives on for centuries. The answer is obvious: Change the law so that lifetime pensions are phased out or converted into time-limited pensions. (Amend the Constitution if you have to.) Those who have paid into social security all their lives but also halted aging at a pre-elderly phase could collect their benefits for, say, 25 years, then go back to work.

5.3 Potential good social consequences

Now let's look at some ways in which the advent of life extension might make the world a better place.

A wiser world

Start with the fact that people often grow in wisdom, perspective, maturity, and patience as they age. If the human race slows or halts aging, we might, as a species, finally grow up. People may have a better grasp of what really matters in life. I have heard people in their 60s, who have been very successful in some career, say that they now believe they were mistaken to value worldly success as much as they did. I regard that as a sign of maturity. The world may grow less materialistic, shallow, and superficial when people have enough time to mature and capitalize on their maturity over a longer period of time.

More concern for posterity

We might, if we live long enough, take the long view more often than we do now. We might care about future generations more, given that we will get to know more of them. We might care more about our long-range impact on the environment a hundred or a thousand years from now. That posterity is us, after all.

More appreciation for many things

We might appreciate our history and our cultural patrimony more than we do now. We might care much more about preserving historical sites and other cultural treasures. As I grow older, I have more appreciation for long-term friendships, kindness, compassion, and other qualities than I did 20 or 30 years ago. I've seen enough ugliness in the world to appreciate the good when I find it. I have a deeper appreciation of first-rate philosophy than I had when I began my studies and a deeper appreciation of good movies and good literature than I had at 35. We learn a lot along the way, and we might learn a lot more along a longer way.

A more informed human race

A population of people who have lived a very long time and tried out many different kinds of work, education, and living arrangements is a very well-informed population. Such people might do a better job of deliberating about social policy and political issues than we do now. As things now stand, most of us know only one profession or a narrow range of occupations and have only a few years or decades of experience. We might make better choices after we have wider and longer experience. Most of us never get the chance to travel extensively for long periods of time or to live overseas and experience another culture at close range. A population where many people have done such things might do a better job of making collective decisions about foreign affairs than we do now.

Better parents

We might do a better job with other things. For example, right now we have to have children while our bodies are young enough to produce them, but imagine having more time to mature, gather economic resources, and get some life experience and *then*, after several decades, tackling parenthood. I'm pretty sure I would have made a mediocre father when I was young. I think I would be much better at it now. Perhaps I'd be better still 50 years from now—or 100 years from now.

New opportunities

Even opportunities may increase in a world where lots of people extend their lives. Yes, those who control existing institutions may lock in their positions, but we would also have more time to create new institutions.

If you have all the time in the world, you can use it to create a new business or institution to compete with existing ones. This may happen more often when people have more time to gain experience, make connections with potential investors, learn from mistakes, and so on. Even if it happens relatively slowly, we may see many more people achieve economic independence sometime during their long lives. Often, we're busy raising families and worried about retirement and wisely decide not to take too many chances. Given more decades or centuries to work with, we might find that we can take chances we can't take so easily now.

A more peaceful world

The world may grow less violent. Violent crimes tend to be committed by the young; gang members, for example, often outgrow their violent tendencies over time. Young people sometimes act as if they think they're immortal or as if their later years mean little to them, but they take fewer chances as they grow older. This may be true of war too. Young people can more easily be persuaded to take physical risks—that's one reason they make the best soldiers. Perhaps an older, life-extended population would be less inclined to go to war, for they would better appreciate how much they lose when they die, and with so much time at stake, they might be far more cautious. We might be less inclined to value the competitive power games that sometimes lead to war. We may have more experience with other cultures and understand them well enough to find other ways to coexist with them. Widespread life extension could bring the world closer to peace.

5.4 Conclusions

K. Most of the concerns about undesirable social consequences, such as pension funding concerns or worries about hierarchies becoming more entrenched, are not trivial but probably not insurmountable. Some of them may well turn out not to be problems at all. In any case, living in a world with these problems is not worse than death. Moreover, the human race has adapted to other, equally profound changes in its history. It's time to do so again. (Section 5.2)

L. There is reason to be optimistic that a world full of people living extended lives will be in many ways a much better world: more mature, more interesting, more stable, wiser, more concerned about the future and the environment, and more at peace. (Section 5.3)

6 Everyone—The Malthusian Threat

6.1 Introduction

In this chapter, we cross a dividing line. Until now we have been discussing concerns about life extension that I consider unproblematic, insignificant, or clearly outweighed by the upside of longer life. Those concerns are interesting and worth discussing, but my judgment is that they are not good reasons to oppose life extension. From this chapter through chapter 11, we will be considering concerns about life extension that I consider very significant. I still think the reasons in favor of developing life extension and making it available outweigh these more significant reasons against doing so, but this is a closer call.

In this chapter, I'll discuss one of the most common worries about life extension: If people live far longer, the earth will become overpopulated, partly because people fail to die on schedule and hang around a lot longer and partly because they have more time to have children. I call this the *Malthusian threat*. I will argue that this is a serious threat and that life extension should be made available only on the condition that those who extend their lives restrict their reproduction enough to avoid a Malthusian crisis. I propose a reproductive policy I call *Forced Choice* to achieve this, and I try to show that it is feasible and just.

Before I begin, I want to express my gratitude and acknowledge a significant debt to Shahin Davoudpour, a doctoral candidate in the school of social sciences at the University of California, Irvine. Mr. Davoudpour is a demographer. He performed all the calculations in this chapter, developed the tables and graphs for the population projections I will discuss, and double-checked my work from a demographer's standpoint. He is effectively the coauthor of this chapter, and I could not possibly have written this without his generous help. Those who are interested in the formula he developed for his calculations will find a detailed explanation of it in section 6.8.

6.2 Will life extension cause a Malthusian crisis?

As we saw in section 1.9, surveys indicate that the Malthusian threat is a major concern for members of the general public. It's also a major concern for ethicists. Peter Singer, for example, asks us to "imagine that we develop and release [a] drug which will slow aging. . . . since people are living twice as long, there will soon be more people than the world can support."[1] Scientists who conduct life extension research run into this concern quite often. According to pathologist Richard Miller, "No one who speaks in public about longevity research goes very far before encountering the widespread belief that research on extending the life span is unethical, because it will create a world with too many old people and not enough room for young folks."[2] Judith Campisi, of the Berkeley National Laboratory, says that after giving a public lecture on aging, "a number of people came up to me and said: 'How dare you do this research? The earth is already being raped by too many people, there is so much garbage, so much pollution.' I was really quite taken aback."[3]

There is reason for serious concern. According to the United Nations Population Division, the world's population was approximately 7.349 billion in 2015, and according to "medium variant" projections (in other words, the middle of a range of projections), it will reach approximately 8.5 billion in 2030, 9.7 billion in 2050, and 11.2 billion in 2100.[4] (According to these projections, most of the growth will occur in less developed countries, particularly in Africa.) Small differences in the fertility rate (the average number of children the average woman has during her life) can pull the projections up or down; in a high-variant projection, the world's population would be 10.9 billion in 2050.

Both because the world's population is so large and because economic development is happening all over the world, we are putting more and more pressure on the world's resources and ecosystem. According to an estimate from the Global Footprint Network, the world is now consuming the equivalent of one and a half earths.[5] In other words, to provide the resources and absorb the waste we are already consuming and producing, without depleting the earth's resources and overloading the planet's capacity to absorb waste, would take a planet half again as big as the earth itself. According to the Network, by the 2030s (less than 15 years from now), we will need two earths. Moreover, roughly 80 percent of the world's population lives on less than \$3,650 a year,[6] but we can expect incomes to rise around the world, putting further pressure on the environment. If this continues, two earths won't be enough.

So we may be facing a Malthusian crisis even without life extension. How much worse will this become if life extension becomes widely available? There are two ways to try to answer that question. First, we can try to predict what will happen. This requires estimating when life extension will become widely available, how large the world's population will be at that point, what percentage of the human race will have access to life extension, how long people with life extension will live, and how many children they are likely to have. However, any estimates of these factors would be almost pure speculation.

There is a better way to approach this. Instead of asking what these demographic trends *will* be, we can ask what they *should* be. In other words, we can ask how much a population would increase given various combinations of life expectancies and fertility rates, look over the projections, and then pick the fertility rate for a given life expectancy that will avoid a Malthusian crisis. This is a matter of simple mathematics and doesn't require trying to predict the future; all we have to do is run some numbers. Think of this as prescriptive demography rather than predictive demography.

Let's start with what it takes to keep the current population steady— neither increasing nor decreasing. This is called the replacement fertility level, and for a population with a normal life expectancy, the figure is an average of 2.1 children per woman (the 0.1 is for infant mortality).[7] What would happen if people who had life extension reproduced at this rate?

To answer this, consider a hypothetical population of one billion people. Everyone in that population uses life extension and lives an extended life. That one billion may be part of a much larger population, most of whom do not use life extension, but right now we're interested in the part of the human race that does, so we'll focus on just that one billion.

I'll make four artificial assumptions about that population: (a) the percentage of women will be 50 percent at all times, (b) no women will experience menopause, (c) infant mortality is zero, and (d) women will have children at average ages specified for each projection (they will have children at various ages of childbirth that will average out to the particular numbers specified). With the possible exception of menopause, these assumptions are almost certainly false. In reality, infant mortality is never zero, the percentage of women in a population is rarely (if ever) a precise 50 percent at any given time, and women will not choose to have children at, for example, precisely ages 25 and 75. However, we don't have the information we need to avoid making assumptions about these matters. When the time comes to limit reproduction, demographers with more information than we have now can work out the details.

To keep this discussion manageable, I'll work with two possible life expectancies: 150 years and 1,000 years. The 150-year life expectancy is chosen because some reputable scientists have suggested that slowing aging enough to produce a 150-year life span is not implausible. The 1,000-year life expectancy is chosen because that's the life expectancy people would have if aging were halted completely and people died only from causes of death that are not related to aging—the far end of what's possible. I will then run projections for each of these life expectancies with various numbers of children ranging from 0.5 children per woman to 3.0 children per mother, at various average ages of childbirth. I focus on the number of children *per woman* not because women should bear the entire brunt of any reproductive policy but simply because they're the childbearers and this is the clearest way to talk about how many children people will have. I will discuss how those limits apply to men when I propose a reproductive policy in the next section.

150-year life expectancy, 2 children per woman at average ages 25 and 75
In the first scenario, the life span is roughly twice as long but the fertility rate is more or less the same as it is now. We will suppose that the average mother's age at the time of the first birth is 25, and her average age at the time of the second birth is 75. These average ages are somewhat arbitrary. I picked the ages of 25 and 75 simply because 25 is close to when many people have their first child now, and 75 is the midpoint of a 150-year life span.

Of course, some people will die sooner than 150 years and some will live longer. Some women may have more than two children, some may have one child, and some may have none. Some will have a child at 18, some at 25, some at 37, some at 59, and so on—the ages of 25 and 75 are merely averages. The total population figures in our projection will not be affected by the fact that these figures are averages. For example, a woman who has her first child at 18 and a woman who has her first child at 32 off-set each other, so we can represent both of them as having their first child at 25 for purposes of calculating the population at a given time. The same thing is true of life span and number of children per woman. (It does, how-ever, matter what the mother's average age at childbirth is. When births are spaced earlier in a woman's life, the population will increase faster, and when they are spaced later in a woman's life, the population will increase more slowly. We'll say more about this later.)

When we run the numbers, it turns out that if people live an average of 150 years and women have an average of one child at an average age of 25 and an average of one child at an average age of 75, *the original population of 1 billion will triple and then stabilize*! Here is the relevant graph (this graph appears again in appendix C as graph 1). The vertical numbers on the left represent the total population in millions, and the horizontal numbers at the bottom represent the generations:

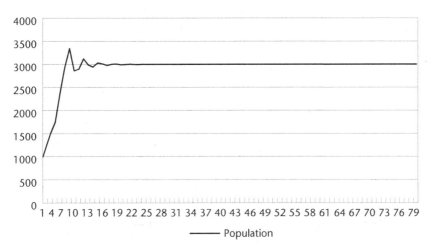

Graph 6.1
Life extension of 150 years, 1.0 child at age 25 and 1.0 child at age 75

This is a surprising result. People are living only a few decades longer and having roughly as many children as they do in many countries now, yet the population triples.

If you want to understand how we got to that conclusion, then we need to start with table 6.1, which illustrates what happens to this population over time. (A longer version of this table appears as table 1 in appendix C.

Table 6.1.
150-year life expectancy, 1 child at average age 25, 1 child at average age 75

Generation	Year	Birth at 25	Birth at 75	Population
0	0	0	0	1000
1	25	500	0	1500
2	50	250	0	1750
3	75	125	500	2375
4	100	312.5	250	2937.5
5	125	281.25	125	3343.75
6	150	203.125	312.5	2859.375
7	175	257.8125	281.25	2898.4375
8	200	269.5313	203.125	3121.0938
9	225	236.3281	257.8125	2990.2344
10	250	247.0703	269.53125	2944.3359
11	275	258.3008	236.328125	3032.7148
12	300	247.3145	247.0703125	3011.4746
13	325	247.1924	258.3007813	2977.9053
14	350	252.7466	247.3144531	3005.3101
15	375	250.0305	247.1923828	3008.3923
16	400	248.6115	252.746582	2993.1488
17	425	250.679	250.0305176	2999.2294
18	450	250.3548	248.6114502	3003.8109
19	475	249.4831	250.6790161	2998.4798
20	500	250.0811	250.3547668	2998.8546

That version carries the calculations out to 80 generations, but this version will stop at 20. There are similar tables in appendix C for all the projections.) The numbers are in millions (1,000 = 1 billion).

Now let's see how table 6.1 was calculated. The other tables in appendix C all work the same way, though with different life expectancies, different fertility rates, and different birth spacings. Some readers will find this tedious, so if you trust my arithmetic, you can skip this discussion

There are five columns from left to right. The column titled "Generation" indicates the generation that comes into existence at that time, starting with our initial generation: generation 0. (Generation 1 is the first generation born *to that group*, not the first group who has life extension. They are the first people *generated*—produced, as it were—by the first group to receive life extension.) "Year" indicates the number of years from the birth of the initial generation (generation 0); we can also use this as the number of years from the advent of life extension in this group. "Birth at 25" indicates the number of people born when the previous generation in the table (the one in the line above) turns 25, and the column titled "Birth at 75" indicates the number of people born when that generation has reached age 75. In table 6.1, the generation that has children at 75 will be three lines above the line where those children appear. For example, the line for generation 3 indicates that 500 million children are born that year—that group of children is born to the women in generation 0, which appears three lines above, for generation 0 has now turned 75.

The column on the far right, titled "Population," indicates the total population—all generations alive by the end of that year. That figure results from adding the previous population figure from the line above to the generation born that year. The generation born that year is the sum of the number under "Birth at 25" and the number under "Birth at 75." For example, the line for generation 3 in table 6.1 includes those born to women who turn 25 that year (125 million children) and those born to women who turn 75 that year (500 million), so generation 3 includes 625 million children.

When a generation dies, that generation is subtracted, of course. For example, in the line for generation 6 in table 6.1, generation 0 has died off, so that one billion people is subtracted from the population to produce the population total for that year.

Now let's walk through the first few lines of table 6.1. We will focus on the total population for each year and see how that yearly population figure is generated, line by line. We will refer to each line by the figure for the "Year" column, using that to indicate the number of years since the birth of generation 0. Generation 0 is the first generation to have a life expectancy of 150 years:

Year 0 population: 1 billion
Half of generation 0 are women. They each have one child when they reach age 25, adding half a billion children (generation 1) to the population:

Year 25 population: 1.5 billion
Generation 1, the first generation born to the original group of people living extended lives, contains 500 million people. In year 50, the women in

generation 1 have one child each when *they* turn 25. This adds 250 million more people (generation 2) to the population:

Year 50 Population: 1.75 billion
In year 75, the women in generation 2 turn 25 and have one child each, adding 125 million people to the population. The women in generation 0 turn 75 and have a second child, adding 500 million children. This produces a total of 625 million people in generation 3. We add that sum to the total population from the previous line:

Year 75 Population: 2.375 billion
In year 100, the women in generation 3 turn 25 and have one child each, adding 312.5 million people to the population, while the women in generation 1 turn 75 and have one child each, adding 250 million people to the population, for a total of 562.5 million people in generation 4:

Year 100 Population: 2.93 billion
In year 125, the women in generation 4 turn 25 and have one child each, adding 281.25 million people to the population, while the women in generation 2 turn 75 and have one child each, adding 125 million people to the population, for a total of 406.25 million people in generation 5:

Year 125 Population: 3.34 billion
This is the year the population peaks. Let's follow this for one more generation.

In year 150, the women in generation 5 turn 25 and have one child each, adding 203.125 million people to the population, while the women in generation 3 turn 75 and each have a second child, adding 312.5 million people to the population, for a total of 515.625 million people in generation 6. That same year, generation 0 reaches the age of 150 and dies off, removing 1 billion people from the population:

Year 150 Population: 2.859375 billion
Table 6.1 carries this projection out for many generations more, but you can see the trend: the population stabilizes at around 3 billion.

The lingering guest problem

The forgoing discussion is meant to show how the population triples even though these people live only roughly 70 years longer than people in developed countries do now and have no more children than the current replacement level. However, this does not quite explain *why* it triples. The answer lies in what I call the *lingering guest* problem.

Imagine that you're holding an open house party for your neighbors. The party begins at 10 o'clock in the morning and continues until 10 o'clock at night—12 hours. Your house can comfortably hold 30 people at any given time. Suppose that each guest stays for an average of one hour and then leaves. Suppose further that they arrive randomly during the day, so the total number of guests is distributed evenly over the 12 hours of your open house party (one-twelfth of them arrive each hour, on average). At any given time, your house will have roughly 30 people in it, and there will be room enough for everyone.

Now suppose your party is more successful than you expected and that all of your guests linger. They arrive randomly during the day (another one-twelfth arriving every hour, on average), but they all stay until the party is over. The house will gradually fill up and up and up, and after a few hours, you'll have several times as many guests as your house can comfortably contain. It will stay that way for the rest of the day.

That is what happens when we live longer. Even if our population of one billion people with 150-year life expectancies has only two children per woman, they are still alive—lingering in the population—several decades longer than people live now. So are their children, who also enjoy 150-year life expectancies, and so are *their* children, and so on. Thus, the generations pile up and hang around instead of clearing out after seven or eight decades, and the population will increase, eventually stabilizing at a much higher level as people finally start leaving the party.

This also explains why the birth spacing—the average ages at which women have children—is important. If women have children earlier in their lives, the population will increase faster, just as your house will get more crowded if all your guests arrive before noon instead of arriving every hour all day. Similarly, if women have children later in their lives or space later births further out (the second child at age 125, for example), then the population will increase more slowly, just as your house will remain uncrowded longer if most guests wait until late afternoon to show up.

Is this a Malthusian crisis?

So the population in this scenario will triple and then stabilize. Is this a Malthusian crisis, or is it something we can live with? That depends on how large a population we are tripling. Initially only some people will have access to life extension, but it would not be safe to assume that life extension will always be unavailable to most people. Suppose that a combination of falling cost and economic growth around the world make it available to anyone who wants it and that most of them choose to get it. The United Nations Population Division medium variant project tells us that the world's population will be approximately 11.2 billion in 2100.[8] If 90 percent of that population chooses extended life, then sometime after 2100, the earth will be carrying more than 33 billion people.

Supporting 30 billion people at a first-world income level would require finding a sustainable and nonpolluting source of energy, solutions to climate change, radical increases in agricultural efficiency, far more complete recycling than we have now, far tighter controls on pollution and refuse dumping in the oceans, an end to overfishing, and so on. Of course, we need to do these things even if the world's population never rises above its present 7.5 billion. If we can do these things at all, perhaps we can do them in a world with a much larger population.

Still, running the risk of increasing the world's population to 30 billion in hopes that technological breakthroughs will arrive in time is one hell of a science experiment to run on the only planet we have, and common sense suggests that the task of saving the world's environment will only be harder if we are coping with population growth at the same time. Moreover, even if we *can* support 30 billion people, such a world would be unpleasantly crowded.

Perhaps things will not get that bad. Perhaps life extension will always be too expensive for most people (though that raises a serious concern about the justice of a world where most people must watch the lucky minority live far longer than anyone else). Perhaps most people will not want it. Perhaps they will have far fewer children.

Perhaps the world's population will be much less than 11.2 billion by the time life extension becomes widely available. Fertility levels are dropping below the replacement level nearly everywhere except in sub-Saharan Africa, and they are falling there too. About 46 percent of the world's population lives in countries with fertility levels below the replacement level, including all of Europe, China, Brazil, Russia, Japan, Vietnam, Iran, Thailand, and the United States, and the worldwide fertility level is expected to drop to or below replacement level by 2095. The International Institute for Applied Systems Analysis (IIASA) in Austria concluded that if the world stabilizes at a fertility level of 1.5—where Europe is right now—the world's population will drop by half by the year 2200, to roughly 3.5 billion. By 2300, according to their projections, it will drop to 1 billion and continue dropping from there.[9]

Let us hope so. However, this period of decline will not happen immediately; the world's population is still increasing right now. Moreover, it's not likely that life extension will be unavailable until 2200—nearly two centuries from now. We are already suffering the beginnings of catastrophic climate change, overfishing, soil depletion, water shortages, and many other Malthusian problems at the current population of just over 7 billion people, and most of those 7 billion are not living at a first-world level. If life extension arrives well before 2200, then all this will be that much worse.

And all of this assumes that women have an average of only *two* children each, despite living many decades longer. If people have an average life expectancy of 150 years and women have an average of *three* children per woman at average ages of 25, 50, and 75, the initial population of one billion grows to nearly seven times its original size in 200 years and grows indefinitely thereafter, reaching a titanic *25 million times* its original size 2,000 years after the advent of life extension! (See table 2 and graph 2 in appendix C.) That third child makes the difference between a population that stabilizes at three times its original size and a population that never stops increasing.

150-year life expectancy with other numbers of children

What if women have only one and a half children each at average ages of 25 and 75 (every other woman has one child at 75)? In that scenario the population increases from 1 billion to 2.656 billion in year 125 but then declines steadily and indefinitely thereafter. By year 250, the total population drops below the original 1 billion; by year 600, it drops below 100 million people; by year 950, it drops below 10 million; and by year 1300, it drops below 1 million—and continues to drop indefinitely. (Table 3 and graph 3, appendix C.)

This is much better than stabilizing at 3.0 times the original population, but it still increases the original population by more than 2.5 times for at least a generation, and it takes 100 years to drift back down to 1.18 billion (roughly the original population), declining further from there. If increasing the world's population to 2.6 times its original size is a Malthusian crisis (and I believe it is), then we would have a temporary Malthusian crisis for the better part of a century.

If women have an average of only one child each at an average age of 25, the population roughly doubles in a little over a century, peaking at 1.968 billion in year 125, and then rapidly drops to below the original 1 billion in the year 150. (It declines by *half* in a mere 25 years.) It continues to drop thereafter. (Table 4 and graph 4, appendix C.) However, even temporarily doubling the world's population is at least a Malthusian threat, even if it does not quite rise to a crisis. We get a similar result if women wait until an average age of 50 to have one child; the population rises to 1.75 times its original size by year 100, drops to less than a quarter of that by year 200, and continues to drop thereafter. (Table 5 and graph 5, appendix C.)

Let's consider an even lower fertility rate: half a child per woman at an average age of 25. Here is the graph (see also table 6 and graph 6, appendix C):

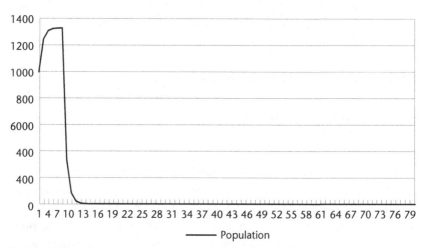

Graph 6.2
Life extension of 150 years, 0.5 child at age 25

Now the population increases by slightly more than 30 percent in 50 years, then levels off, peaking at 1.333 billion in the year 125. A generation later, in the year 150, the population has plummeted to 333 million and declines dramatically from there on—to less than 1 million by year 275. It then fluctuates indefinitely between a few million and a few hundred thousand but never reaches even 20 million from then on. Obviously, we would need to raise the rate of reproduction soon after the population peaks, but to achieve this result, the initial limit must be *0.5 children for 125 years*. (We get a similar result if women wait until an average age of 50 to have 0.5 children each, so delaying the age of childbirth won't make much difference. See table 7 and graph 7, appendix C.)

Would allowing the world's population to increase by one-third for 125 years be an intolerable Malthusian crisis? My personal judgment— for whatever that's worth—is that *this* is a tolerable increase, particularly since it lasts for only a century.

Bear in mind that the 0.5-child limit applies only to those who extend their lives. Moreover, this limit applies only for 125 years; after that, everyone can have up to 2.1 children per woman (or whatever the replacement level must be for the entire population to stabilize). The downside, of course, is that every other woman in the population of people living extended lives will not have a child at all. (This limit applies to men as well. I'll explain how that works in section 6.3.)

I conclude that a 0.5-child limit for 125 years is a plausible initial target birthrate if life extension provides a life expectancy of 150 years and everyone in the world has access to life extension. This may not be the actual limit when the time comes, of course. If the life expectancy is less or not everyone has access, the limit could be correspondingly higher. We can't pin down exactly what the target fertility level should be without knowing more about the extended life expectancy and the number of people who would have access to life extension, but we do know that if life extension gives us at least several extra decades and becomes widely available, then we must temporarily limit reproduction to significantly less than two children per woman for those who use life extension.

1,000-year life expectancy

Of course, it's not safe to assume that we'll never be able to extend the human life expectancy past 150 years. If it's possible to slow aging, it may be possible to slow it to a halt. As I explained in section 1.7, if we halted aging and the current rate of accidents and violence were unchanged, our life expectancy would be around 1,000 years. Some people would live much longer, some would die much sooner, but the various life spans would average out to approximately 1,000 years. What must our reproductive policy be like in order to keep life extension from making our Malthusian problems worse in *that* world?

Once again, let's start by supposing that each woman has two children—the reproductive limit necessary to prevent excessive population growth in the absence of life extension. Suppose further that each woman has one child at age 25 and one child at age 500. Specifying age 500 as the age for having a second child is arbitrary; I chose it simply because it's the midway point. Bear in mind that if births occur earlier in life, the population grows much larger and faster and that if they occur later in life, the population will grow more slowly and less.

Here is the relevant graph (this graph appears as graph 8 in appendix C; see also table 8, appendix C):

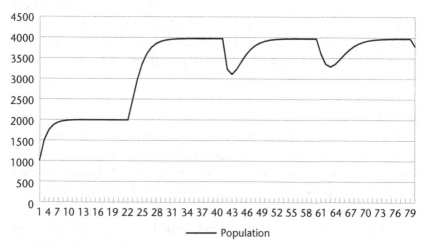

Graph 6.3

Life extension of 1000 years, 1.0 child at age 25 and 1.0 child at age 500

This population increases from 1 billion to nearly 2 billion very quickly—in just 75 years. The population then levels off at roughly 2 billion for the next 400 years. From year 475 to year 600, it increases to nearly 4 billion, stays at roughly that level until year 975, and then fluctuates between 3 and 4 billion for centuries, eventually stabilizing at around 4 billion. Notice that this is only 1 billion more than the eventual population in the scenario where the life expectancy is only 150 years, even though people are living several times longer in this scenario. The Malthusian threat posed by radical, 1,000-year life extension may not be much worse than the one posed by moderate, 150-year life extension!

Still, we had serious concerns about whether increasing the population threefold is tolerable, and those concerns are even more serious when we consider increasing the population fourfold. And that's just for starters. If this population has *three* children per woman (at average mother's ages of 25, 50, and 500), the population reaches nearly *seven* times its original size by year 225 and continues increasing thereafter, reaching a staggering *1,700 times* its original size by year 2,000. (Table 9 and graph 9, appendix C.)[10]

What if women have an average of one and a half children at average ages of 25 and 500 (half the women have a second child at 500)? As we can see from table 10 and graph 10 in appendix C, the population roughly doubles by year 100, rises to 3 billion around year 700, then declines to around 2 billion around year 1,000 and then slowly declines for many centuries thereafter. However, the population still triples for most of a thousand years. This too is a Malthusian crisis.

If women have an average of only *one* child at an average age of 25, the population roughly doubles in the first 100 years and then remains at that level for another 875 years. It then declines over a period of many centuries. (Table 11 and graph 11, appendix C.) I don't know about you, but doubling the world's population for 875 years sounds like a Malthusian crisis to me.

Now consider what happens if women have an average of half a child each at an average age of 25 (half the women never have a child at all, and the other half have no additional children for a thousand years). Here is the relevant graph (see also table 12 in appendix C, where this graph appears again as graph 12):

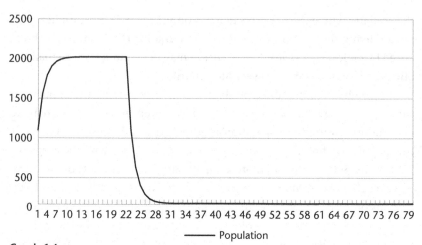

Graph 6.4
Life extension of 1000 years, 1.0 child at age 50

In this scenario, the population rises to 1.250 billion in year 25, 1.312 billion in year 50, and peaks at 1.333 billion in year 125. The population then stabilizes at 1.333 for 850 years (to year 975) and then starts dropping sharply and continuously from then on. If women postpone that half of a child until age 50, we get a similar result: the population rises to 1.250 billion in year 50, 1.312 billion in year 100, and 1.333 billion in year 250. (See table 13 and graph 13, appendix C.) That population levels off until year 950 and then begins to decline continuously and sharply.

Is increasing that world's population by one-third for several centuries a tolerable increase? Let's put this in perspective. If the growth estimates we considered earlier are correct, by 2100, the current world population of roughly 7.5 billion will increase by approximately 45 percent according to the United Nations Population Division and approximately 25 percent according to the Institute for Applied Systems Analysis (IIASA) in Austria. We are likely to see the world's current population increase by one-third even in the absence of life extension, and the spread in estimates is fairly large: 1.9 billion (9 billion according to the Austrians, 10.9 according to the UN).

My judgment is that if these increases are tolerable, then a one-third increase on top of that is not *in*tolerable. If we can find renewable sources of energy, more efficient ways to produce food, better ways to dispose of waste, and solutions to climate change for a population numbering in the billions, then we should be able to find those things for a population one-third larger than that. (Of course, this assumes we *can* achieve all those things. If we can't, then we may be doomed in any event.)

So I reach the same conclusion: if life extension halts aging completely and the entire world gets a life expectancy of 1,000 years, a plausible target fertility level is 0.5 children per woman at an average age of 25 or above—at least for 1,000 years or so. Of course, that 0.5 figure is only one of many plausible estimates, and it might go up if childbirth is delayed. My main point is to demonstrate that a population with life extension reproducing at anything close to current fertility levels poses a very serious Malthusian risk.

6.3 A policy to prevent a Malthusian crisis: Forced Choice

Suppose we want to design a policy that enforces a target birthrate of 0.5 children per woman. (I pick that target just for illustration; what I say in this section doesn't turn on that particular target.) I'll call this policy "Forced Choice," for it forces people to choose between extended life and having more children than the target birthrate—those who choose to have more children are denied life extension.

Here is one way Forced Choice could work. We are supposing that those who extend their lives can have an average of only 0.5 children per woman at an average age of 25 at childbirth. In effect, half of all couples have one child, and half have none. This policy is imposed on men too. (To be clear, it is applied on an individual basis, not on couples as units.) So divide one child among four people (one child for every two couples): each person gets a 25 percent chance of having a child. That is, one person out of every four gets a chance to reproduce. To decide who gets that chance, those who wish to extend their lives must enter a reproductive lottery (or be denied access to life extension). The lottery applies to both men and women and any other genders, and it applies in the same way to all of them. If you win the right to reproduce, anyone you reproduce with can do so as well, even if they lost the lottery. In other words, if you win the lottery, you don't have to look around for another lottery winner in order to exercise your right. You can share your good fortune with a partner who lost the lottery but won you.

If you have had no children yet and you enter the lottery, you have a 25 percent chance of winning the right to reproduce. If you have *already* had one child and you then enter the lottery, the trade-off is reversed, in a way: you enter the lottery with a 25 percent chance of winning the right to extend your life. Thus, having one child reduces your chances of being allowed to extend your life from 100 percent to 25 percent. You'll want to consider that first child very carefully; it could be a fatal choice.

If you have had *two* children already, then you are forbidden to have life extension: your odds of extended life are now zero.

The limit of 0.5 children per woman was based on the expectation that people who have a child would do so at age 25. However, if they wait until a later age, then the lingering guest phenomenon is lessened, and we could keep the population within a tolerable limit with a higher reproductive limit. Of course, Forced Choice is not meant to tell anyone they *must* have a child by a particular age, and this opens up another possibility. Delaying the average age of childbirth slows the population increase to some extent, so your odds of winning the right to reproduce could be correspondingly higher the longer you wait to enter the lottery (linked to whatever higher rate of childbirth at that average age of childbirth will produce the same population increase). And here is another possibility: if you lose the lottery when seeking to reproduce at age 25, you might subtract your odds at age 25 (25 percent from your higher odds at some later age) and enter the lottery again at that later age with a percentile chance of reproducing equivalent to the difference between 25 percent and whatever higher odds you would have had if you had waited until that later age to enter the lottery the first time. We can even imagine repeating this several times in a lifetime with various odds of success, though the more often one does this, the lower the odds would be each time one enters the lottery.[11]

Those who extend their lives may, of course, have children with people who do not extend their lives. However, the reproductive limits still apply to the person who lives an extended life. For example, if you have extended your life and you want to have a child with someone who has not done so, your partner is not limited, but you are, and the two of you can have a child only if *you* can have a child. If your partner can but you can't, then your partner can have a child only with someone else who has the right to reproduce. (You might, presumably, be able to adopt the child of that union.) Otherwise, the limit on reproduction would be ineffective, for anyone living an extended life would merely need to reproduce with someone who is not. To see this, suppose our entire population of one billion people living extended lives had children with a billion other people whose lives are not extended: the result is a billion children when the target calls for 250 million children (0.25 children per person).

These limits do not apply to adoption, except insofar as the child comes from parents living extended lives.

If you are living an extended life and your child dies, Forced Choice allows you to have another child to replace the one who died. The reproductive limit is based on the number of children, not the number of childbirths.

This works for same-sex couples too. Each adult gets a 25 percent chance regardless of gender identity or sexual orientation. If you're gay and you win the right to reproduce, you and your same-sex partner can have a child, just as a heterosexual couple would. This also works for people who are infertile: each adult gets a 0.25 chance even if they are biologically unable to reproduce for some reason. An infertile person who wins the lottery can then use assisted reproductive technology and egg and/or sperm donors (or even cloning) to exercise that right and have a child. If that person is unable to reproduce even with the help of such technology, then they might be allowed to sell or give that right to someone else who can use it, somewhat like a carbon credit. We might allow this also for those who win the lottery and decide not to use their right to reproduce—that way it won't go to waste.

If you've already extended your life and you then have a child without having won the reproductive right to do so, your life-extending treatments are terminated and you resume aging at a normal pace. (Remember that life-extending treatments are likely to require periodic or ongoing interventions of some kind, such as drugs or stem cell treatments. This is not likely to be akin to a vaccine you take only once in your life. Thus, it won't be hard to cut people off.) Your reproduction is now no longer limited (assuming your society has not imposed some other limit on reproduction that applies to everyone), but you will resume aging.

In other words, your decision to extend your life and limit your reproduction is reversible. If you extend your life and you later wish you had reproduced instead (or wish you had more children than you were allowed to), you simply reverse that decision and revert to normal aging, just as if you had never extended your life in the first place. You can then have more children. Of course, a decision to have two children and then extend your life is *not* reversible.

You can make these choices at any time during your life. You reduce your choice only if you have one child before you request life extension, and you lose it only if you have two or more children before you request life extension. The timing of that request is up to you.

What if lots of people extend their lives and then, after many decades or centuries, they decide to discontinue their life-extending treatments and *then* have an extra child? In that event the population of those who live extended lives might get too large due to the lingering guest problem (having one child and living 280 years will create more population pressure than having one child and living 80 years). In that event, the target birthrate might have to be lowered to compensate for this.

Note that this is not a complete population policy, for I've said nothing about limiting the reproductive choices of those who cannot or choose not to get life extension. Thus, Forced Choice may not prevent a population crisis; we might overpopulate the earth even without life extension. We may need other policies to prevent that. Forced Choice is meant merely to ensure that life extension does not make that crisis worse.

6.4 Practical problems with Forced Choice

How do we know when people have had life extension but exceeded their reproductive limit? In many cases, this is straightforward. Societies around the world would have to keep track of birth records to see who has had children and how many children each person has had. Life extension clinics would have to be licensed and supervised, and each person who comes in would have to be identified (perhaps with a pinprick blood sample and a DNA chip[12]) and their reproductive records called up. If that person has exceeded his or her reproductive limit, then the penalties kick in.

Of course, some people will try to hide their children and keep society from finding out about their children's birth and parentage. However, it should not be difficult to use genetic testing (again, with a pinprick of blood and a DNA chip) to read every child's genome. That genome can then be compared with all others in the system and the biological parents (or at least sperm or egg donors) identified. There would be no penalty for the child, obviously, but if a child is traced to a parent who has exceeded his or her reproductive limit, then the parent is penalized. If both parents have exceeded their limits, then both parents are penalized. One would have to hide a child from the public schools, the healthcare system, and all other authorities indefinitely to do this, and while that's possible, it's sufficiently difficult that few people will succeed indefinitely.

Some people might seek to avoid Forced Choice in another way. Instead of hiding their children, they might try to hide the fact that they are getting life extension. For example, they might fly to some overseas clinic operating beyond the reach of the authorities or have a back-alley operation at home and get their life-extending treatments from a source that doesn't care about their reproductive history.

There are a couple of ways to deal with this. One way is to try to find and shut down any illicit operations. Another way is to police the user. We could take a pin-prick blood sample and a genetic reading with a DNA chip, check that against the birth records, and thereby see how old someone is. Anyone above a certain age is obviously using life extension. That genome can then be compared with the genetic records for all children, and if any of the children turn out to be a match with that adult, then we know how many children that person has had.

These methods of enforcement are not airtight (law enforcement never is), but they don't have to be. Forced Choice does not have to be enforced in absolutely all cases in order to be effective. It can be effective at staving off a Malthusian crisis provided it's enforced in the great majority of cases. Moreover, Forced Choice need not be entirely effective at first. All of these details can be adjusted and revised as needed for some decades, for a Malthusian crisis would develop slowly enough for us to watch trends and see what works.

It may be cumbersome and difficult to develop this system and enforce it. Is there a better alternative? We could simply let people reproduce at will and see what happens (a policy of Free Choice), but that may turn out badly for the human race. We could also try to ban life extension altogether (a policy of Prohibition concerning access). However, banning life extension may be just as difficult as controlling the reproduction of those who extend their lives and require nearly as much record-keeping, for we would still have to keep track of birth records and run genetic tests on anyone who has lived past a certain age. Moreover, banning life extension altogether may be politically more controversial and unpopular than controlling the reproduction of those who extend their lives. Prohibiting life extension denies decades or centuries of life to people who want it, while Forced Choice merely limits how many children they have.

There really is no better alternative to limiting the reproduction of those who extend their lives, at least for a few generations.

6.5 Reproductive ethics and Forced Choice

Forced Choice raises some difficult issues concerning the ethics of individual reproductive choice. I will offer some fairly brief responses to these issues. We may not be able to analyze them fully until after Forced Choice is implemented, when such cases arise in practice.

To begin with, Forced Choice assumes that people *choose* between extended life and having more children, but what about those who don't reproduce voluntarily? There are many versions of this case, including people who are raped, people who have sex with someone who is too young or otherwise not competent to make that choice, people who use forms of birth control that fail for some reason (especially if they have moral objections to abortion—the birth control of last resort), and people who had good reason to believe they were not at risk of becoming a mother or father (perhaps due to infertility) but became one anyway (again, especially if they have moral objections to abortion).

Each of these cases concerns whether or not someone voluntarily became a parent. Implementing Forced Choice will require carefully thinking through such cases and deciding when someone's reproductive choice was truly voluntary, competent, informed, and so on. For example, a clear case of rape is a clear case where the victim should be treated as if her child had never come into existence so far as access to life extension is concerned. A more subtle, ambiguous case involving, for example, someone of dubious competence, might be harder to analyze. However each case is decided, we can say this much: if we decide that the mother or father did voluntarily reproduce, then that birth counts under Forced Choice, and if we decide that she or he did not, then that birth does not. Many of these cases might have to be settled by the courts.

What if you lose the reproductive lottery and you don't want to forgo extended life? Presumably you're free to try to find a mate who won the lottery. However, this might make it much harder to create and sustain a committed relationship and give a huge advantage to the winners of the reproductive lottery. As I mentioned earlier, some winners in the reproductive lottery might choose not to exercise their right, instead selling or giving it to someone else, like carbon credits. To that extent, reproduction would be yet another privilege of the rich. If that concerns us, we might ban selling reproductive rights to prevent such inequality, just as we ban the sale of organs. My judgment is that these problems are just part of the price we would have to pay to live in a world with life extension and avoid a Malthusian crisis, but they are unquestionably a downside to that world.

Of course, adoption is still possible under Forced Choice. Presumably the great majority of children available for adoption would come from people who are not extending their lives. Does this create an incentive for Have-nots to produce babies for the adoption market? If so, society might have to ban any form of compensation to those who produce such children, or impose limits on such adoptions, or lower the target birthrate to compensate for increased reproduction by Have-nots.

What about those who had good reason to believe they would never have access to life extension (they were born poor), so they went ahead and had children, but later found that they could have life extension after all, perhaps because they suddenly came into money? Their choice to reproduce was made under conditions where they had reason to believe extended life was not an option. Is it fair to deny them life extension later on just because they didn't know they would have access one day?

Frankly, I'm not sure how to handle such cases. We could simply take a hard line and say that people must weigh those odds when they choose to have children and take their chances. We could also say that being denied extended life because you are poor is no worse than being denied extended life because you mistakenly thought you would always be poor *and you had children*.

A better approach might be this: those who came into money later in life could be allowed to appeal the denial of extended life and try to prove to some tribunal that they had good reason to believe they would never be able to pay for life extension. People who won millions of dollars in a monetary lottery would be able to meet that standard quite easily, while those who simply succeeded in business (due to the foreseeable effects of hard work) might have a more difficult time meeting it.

In the end, it's impossible to entirely avoid inequity in the details of a reproductive policy like Forced Choice. The price of making life extension available and avoiding a Malthusian crisis may simply be to tolerate the partial injustice of any real world version of Forced Choice.

6.6 Is Forced Choice an oppressive government intrusion into private reproductive choices?

Forced Choice may sound a lot like China's one-child policy, and it might trigger an objection sometimes leveled at that policy. You might object that this is an unacceptable intrusion into the personal affairs of citizens and that governments should not tell people how many children to have. However, that objection is not valid when applied to Forced Choice. Forced Choice is very different from China's one-child policy.

First, Forced Choice doesn't set limits on how many children people can have when they don't extend their lives. (We may want to limit reproduction within that period for reasons having nothing to do with life extension, but that's another matter.) Second, Forced Choice doesn't apply to everyone, aside from taking genetic samples for record-keeping. The limits it imposes apply only to those who choose extended life, and by choosing extended life, they can be said to have consented to certain limits on their behavior. Third, Forced Choice is meant to prevent harm to people who don't want life extension (the Will-nots) or can't get it (the Have-nots). If there is a right of reproductive freedom (as an instance of the right to self-determination in general), that right—like the right to self-determination generally—is subject to the harm principle. It's generally recognized that you have a right to determine your own life (including your reproductive life) but that you may not do so in ways that harm other people. Fourth, if there is something biologically unnatural about living without aging (and by "unnatural" I don't mean "therefore bad"), then we should accept that such a life might be bound by biologically unnatural limits on reproductive behavior. A new life may require new ways of living, and seen from that perspective, Forced Choice should not seem unduly restrictive or intrusive.

6.7 What if some countries impose Forced Choice and others do not?

What if not all countries impose Forced Choice? Suppose country A makes life extension widely available but imposes no limits on reproduction. We live in country B, and we're deciding whether to impose Forced Choice.

We see what's happening over in country A, and many of our citizens complain at living under constraints that aren't applied to the citizens of country A.

Does the fact that we (in country B) see country A behaving recklessly in its population policy give us a reason to be reckless too? Surely not. The fact that someone else is contributing to a problem is not a reason for you to make things worse too. It may seem like it is if everyone else is ignoring the problem and you are the only one doing something about it, just as you might give up trying to recycle your trash if everyone around you fails to recycle, but that's not how we usually think about moral duties. We tend to think that our moral duties are not contingent on everyone else performing that duty too, even if we must all perform that duty in order to avoid some widespread harm.

To see why, consider another response to country A: we might decide that A's failure is a reason for us to control our own population even *more* severely to offset the harm A is doing. In other words, instead of instituting Forced Choice, which allows limited reproduction, we might institute Prohibition either for reproduction or for access to life extension, or both.

Do we have a duty to do this? We don't generally think so in other situations. We don't, for example, think that you have a duty to do *more* than your share to clean up the local environment just because others do less than their share. Your duty is to do your share—period. You cannot have a duty to do more than your share, for that amounts to a duty to do more than your duty. Imposing a harsher duty penalizes those who do their share, and that seems unreasonable. Similarly, country B has no duty to inhibit or prohibit access to life extension or impose more draconian limits on population than Forced Choice calls for just because there are other countries that don't do enough. Forced Choice is enough, regardless of what other countries may be doing.

If your share is not increased by someone else's dereliction, then your share is also not decreased by someone else's dereliction. Whether someone else performs their duty is morally irrelevant here. Therefore, the fact that country A does not impose Forced Choice on its citizens is not a reason for country B to allow those living extended lives to reproduce without limit.

I'm not dismissing concerns about whether an *individual's* moral duty is sometimes contingent on others performing their duty. A duty to obey the law, for example, might be permissibly ignored in a society where everyone else ignores the law, especially when obeying the law imposes some cost on those who obey it. However, we're not talking about an individual's duty to limit his or her reproduction. (In this discussion I'm neither claiming nor denying that such a duty exists.) We're talking about whether an entire country has such a duty. When an entire country fails to limit its population, especially if the country is large, that dereliction has consequences that are significantly harmful for the entire world in the long run. (Similarly, an individual may have a duty to obey the law in a very small society where everyone else ignores the law if that individual's disobedience imposes significant harm on everyone else.) I'm not suggesting that potential harm is the only source of a duty, but it can override other reasons to suppose a duty has been suspended.

Of course, getting every country (or at least most of them) to institute a responsible reproductive policy for those citizens who extend their lives will be necessary to avoiding a Malthusian crisis, and that may be very difficult. However, that's a problem in enforcing our duty to control reproduction, not a reason to think we have no such duty.

That said, there is some reason for guarded optimism that all or most countries will eventually impose some reasonable limits on population, for the world is starting to do something like that concerning climate change. By 2016, 195 countries had signed a pact popularly known as the Paris Climate Agreement, promising to do what is necessary to keep the average global temperature from rising more than 2° Celsius over preindustrial levels, provided that enough of the countries signing that pact ratified the agreement. In 2017, Donald Trump declared that the US would not abide by it, yet several American states and more than three hundred American cities have declared that *they* will, so not all hope is lost. Perhaps we can achieve something similar for limiting reproduction worldwide. In fact, it may be even easier to limit reproduction worldwide, for unlike preventing climate change, Forced Choice doesn't require technologies that have not been developed, radical changes in our economy, or reducing our material consumption. Moreover, the population trends that make Forced Choice necessary will unfold slowly, generation by generation, while the world's climate may suffer serious deterioration as soon as a century from now. Therefore, we have more time to institute Forced Choice than we have to prevent climate change.

6.8 The demographic formula used in this chapter

The formula for the calculations behind the tables and graphs in this chapter and in Appendix C was developed by Shahin Davoudpour, who also co-authored this section. For ease of exposition we did not go into the formula earlier, but for those who are interested, here it is:

$$P_t + P_{t-1} + \Sigma_{g=1}^{n}(\Phi_g P_g \beta_g)t - \Sigma_{g=1}^{n}(\Theta_g)t$$

We want to know what happens to a given population over a span of time ending at time t. P_t is the total population at time t. P_{t-1} is the initial population at an earlier time. Φ_g is the gender ratio of a particular generation g (for simplicity we set the ratio at 50/50 in all our calculations). β_g is the number of children allowed per woman in generation g at time t. θ_g is the number of deaths for generation g at time t. Finally, we must calculate for all generations during the span of time we considering. Hence, Σ indicates that the sum of the multiplication of the three variables in the parentheses for each generation of women having children at that time, which are between 1 and n (1,n), must be added to the sum for every other generation of women having children at that time. (In other words, more than one generation of women may be having babies at that time; we must run the numbers for all generations having babies at that time and add them up.)

Suppose, for example, that we want to know how much an initial population of one billion will grow by the birth of the third generation, fifty years after that one billion begins using life extension, where people have an average life expectancy of 150 years, the gender ratio is 50/50, and women have an average of one child at age 25 and .5 children at age 50. P_0 (P_{t-1}, t is 1, hence P_0) is the initial population. Variable g is 1 since it is the first generation, Φ_1 is .5 (the percentage of women in that population of 1 billion [P_1] is 50%), and β_1 is 1 child per woman for the generation of women who have their first child in that year (in other words, those who just turned 25 years of age). We get a total of 500 million children. Now the population of the world is 1.5 billion.

Twenty-five years later, the women in g_1 turn 50; they are now entitled to have an additional .5 children each. We run the numbers again (except that this time β_1 is .5 children per woman) and get another 250 million children. At the same time, 500 million people turn 25 (g_2), and the women in that group have an average of 1 child each. We run the numbers again with β_2 of 1, and get an additional 250 million births. Finally, we add the initial 1.5 billion to 250 million (those who are 50 years old) and 250 million (those who are 25 years old), for a total of 2 billion people at time t, 50 years after life extension became available to this population.

6.9 Conclusion

M. There is good reason to believe that making life extension widely available will pose a serious risk of a Malthusian crisis of overpopulation, pollution, and resource shortages. To avoid this, it is very likely that we will have to significantly limit the reproduction of those who extend their lives. (Section 6.2)

N. A policy of Forced Choice, which requires people to choose between extending their lives and having as many children as they wish, is feasible and just. (Sections 6.3, 6.4, 6.5, 6.6, and 6.7)

7   The Have-nots—Distress and the Death Burden

7.1 Introduction

In the last two chapters we looked at harms that affect more than one group. Those harms, if they occur, will affect the Have-nots, but they will also affect the Haves and the Will-nots. In this chapter we will look at possible harms that are unique to the Have-nots.[1]

The problem starts here: If you die at 90 in a world where no one lives much past 100, your death is not so bad. Death is always a great loss, but you lived about as long as it's possible to live, and you had a good run. If, however, you live in a world where life extension is available to the rich but *you* must age normally and die at 90, your death seems worse. Ninety years no longer seems like enough time when others get a lot more time, even though you get just as much time as you would in a world without life extension. Dying at 90 under those conditions seems much worse than dying at 90 in a world where no one lives much past a century.

If it's worse to die at 90 in a world where some people can afford life extension but you can't, then introducing life extension into the world seems to make the Have-nots worse off in just that way. In other words, introducing life extension into the world seems to make billions of people worse off even if their lives are no shorter. However, it's not immediately obvious *why* they are worse off. After all, their lives are no shorter. We don't usually think that we become worse off just because someone else becomes better off: if my neighbor wins the lottery, I do not thereby become poor. What, exactly, is the harm here?

7.2 Distress

One obvious answer is that many or most Have-nots will suffer feelings we'd all rather avoid: horror, despair, depression, anger, resentment, and other feelings so unpleasant that experiencing them counts as a harm in its own right. Let's refer to the entire set of these feelings as "distress." Introducing life extension without making it available to everyone seems to harm billions of people by making them seriously distressed.

It can be hard to disentangle distress from questions of justice: the Have-nots may feel distressed because it seems unfair that others get life extension and they don't, and they may be right that it's unfair. We'll discuss justice in the next three chapters, but in this chapter, I want to focus on distress by itself. To distinguish distress questions from justice questions, let's imagine a case that leaves justice out of this: Imagine that it's impossible to provide life extension for a portion of the human race because those people are physiologically unable to benefit from the treatment. Life extension works only for those who have the right genes; these people have the wrong genes, and there's no genetic intervention that affects those genes. They are genetic Have-nots. Because it's impossible to provide them with life extension, the fact that they don't get it is not evidence of injustice—just bad luck.[2] Justice aside, what is true about the distress felt by genetic Have-nots should be true of the distress felt by those who are Have-nots because they lack the money to buy life extension. In other words, these imaginary genetic Have-nots will help us think about distress in a context where issues of justice do not arise, so we can isolate questions of distress from questions of justice.

If we set aside issues of justice, what else do the Have-nots in our case have to be distressed about? What about the potential Malthusian consequences and other possible bad social consequences we talked about in the last two chapters? Those harms may be significant, but it's still worth asking whether distress alone counts as a morally significant harm in its own right, even leaving those possible harms aside. For one thing, the Malthusian harms and other bad social consequences might be avoided or turn out to be insignificant. For another, Malthusian and other bad social consequences won't happen right away; some of them may take generations to fully materialize. However, Have-nots will probably suffer serious distress as soon as they hear about the first Silicon Valley billionaires who get life extension.

Perhaps the Have-nots are distressed because the Haves get extended life and the Have-nots don't. The mere fact that the Haves are better off leaves the Have-nots comparatively worse off, even if they are not worse off in an absolute sense. If, for example, your income remains the same but other people get raises, you're worse off compared to them but not worse off in any other way. But is being comparatively worse off a harm? We almost never call it a harm. If your neighbor wins the lottery, you're now comparatively worse off than he is, but we don't say you were harmed by this. If your house was burglarized and you lost property, you are worse off than your neighbor who was not burglarized, but we don't think the mere fact that he now has more property than you is a harm to you apart from what the burglar did.

Refusing to count being comparatively worse off (without being absolutely worse off) as a harm not only fits our moral judgments but also makes sense. Recognizing every instance of being worse off than other people as a harm would expand the right against harm so massively that no one could do much of anything without running the risk that they are harming someone else. A set of moral rules that contained such a right would not promote human flourishing. Moreover, prohibiting anything that makes some people better off than other people would amount to an extremely radical kind of egalitarianism—something no egalitarian would support.[3]

If there are no purely comparative harms, then we're talking about distress in the absence of harm (at least before Malthusian and other collective harms materialize). Distress without corresponding harm is psychologically plausible, of course, but it's morally puzzling. If the Have-nots are distressed but not otherwise harmed, what, if anything, should we do about their distress? Can distress without a harm to be distressed *about* count as a harm all by itself?

Let's address those questions by considering this case: A homeowner feels very distressed when he learns that the couple who bought the house next door is gay. He's beside himself over the fact that the two women living next door sleep together and have sex. Should we take that distress as seriously as we would if the homeowner felt that amount of distress because the couple next door held loud parties and kept him up every night? Being kept awake every night is a real harm to the homeowner, and he would be justified in feeling distressed about it. He's not justified in feeling distressed because the people next door are gay. Therefore, his distress at their behavior does not have the moral weight it would if they did some real harm to him.

Of course, the homeowner may be unable to control his reaction. In that case his distress might have *some* moral weight. We might then take his distress seriously in certain ways: We would avoid goading him about his neighbors, try to be patient and diplomatic when we talk to him about his neighbors, recognize that he can't help this, and so on. Moreover, if we could help him eradicate his distress by coming to see that his neighbors are good people who are harming no one, we should do so—not just because his neighbors deserve this but also because the homeowner himself will be better off without that distress. His distress is not devoid of moral significance. Nonetheless, we would still give his distress less moral weight. His neighbors have a moral duty to end their parties by ten o'clock, but they don't have a duty to sleep separately.

Now transfer this analysis to life extension. Although the Have-nots will eventually suffer other harms, let's stipulate for purposes of discussion that at least early on, the Have-nots are not suffering any harm apart from the distress, just to isolate that issue. Should we do anything about their distress as a harm in its own right? We may have a moral duty to be compassionate, to try to ease their distress in some ways, and so on, especially if they can't help feeling distressed and their distress is severe. Nonetheless, their distress is not as strong a reason to inhibit or prohibit life extension, all else being equal, as it would be if they suffered a harm apart from their distress.

Let me sharpen this a bit. I don't simply mean that their distress is a reason of X weight, and the other harm is a reason of Y weight and that X + Y is a greater total weight of reasons to inhibit life extension than the moral weight of X alone. Anyone would agree with that. What I mean is, if there is no other harm of Y weight, then their distress does not have X weight in the first place—it has less than X weight: X − n. That's not to say that their distress has no moral weight at all or that Haves can disregard the Have-nots' distress entirely. Even if the moral weight of their distress is X − n, X − n weighs *something*. The Haves have a duty to be sympathetic, to avoid rubbing it in, and perhaps even a duty to compensate the Have-nots to some extent for that distress. If, however, the Have-nots had another harm to be distressed about, then the moral weight of the Have-not distress would be X, not X − n.

Whether their distress, whether it has moral weight X − n or moral weight X, is a sufficient reason to inhibit or prohibit life extension is another question. Nothing I've said so far implies that distress without another harm could not, by itself, be a sufficiently large harm to make it wrong to develop life extension—I leave that question open for now. (I'll later argue that life extension should not be inhibited or prohibited at all, even when we consider all the possible harms, so you know where I'm going to come down on this.) My conclusion here is simply that their distress by itself has less moral weight than it would have if it were accompanied by some other harm for them to be distressed about.

However, even before Malthusian consequences or other harms have a chance to materialize, and even apart from issues of equality and justice, there *is* a separate harm the Have-nots will suffer, and this other harm will arrive as soon as life extension itself arrives. I call it the death burden.

7.3 The death burden

The severity of death

Some deaths are worse than others. Dying at 9 seems far worse than dying at 90. All else being equal, the younger you are when you die, the worse your death is, for the number of years you might have lived is larger. This intuition is, I suspect, so universal and so strong that we can propose it as a partial account of how bad it is to die. It's a counterfactual account, for it turns on what would have happened if (counter to the facts) you had not died then. When we talk about how bad a particular death is for the one who dies, let's call that the *severity of death*. When we talk about how bad it is to die at a particular age (all else being equal), let's call that the *temporal death burden*, or *death burden* for short. I propose a counterfactual measure of the severity of death: your death is worse, all else being equal, the more years you would have lived had you not died when you did, and the number of potential years you missed out on is your death burden. (I'm not the first to offer such an account. Many philosophers have done so, particularly Jeff McMahan, who explores the counterfactual account of the severity of death in greater depth and detail than anyone else, though he doesn't call it that.[4])

The severity of death is not what philosophers call the *harm of death*. The problem of the harm of death concerns this question: If you no longer exist once you die, how can death harm you? (The problem is that it's hard to see how anything can happen to you—including harm—if you no longer exist.) The severity issue is a different problem. It concerns *how* bad death is, while the harm issue concerns how it can be bad at all. They sound similar, for I'm proposing a counterfactual measure of the severity of death, and the most popular account of death's harm is that death harms us by depriving us of potential years of life—roughly the same counterfactual. However, it's possible to hold a counterfactual view of the severity of death while denying the deprivation account, for we can distinguish between using the number of years you would have lived as an *index* of harm and using them in an *explanation* of harm. This is important because not everyone accepts the deprivation account of the harm of death. Stephen Blatti, for example, is a nondeprivationist who argues that part of death's harm consists in the fact that death constrains our autonomy; in plain English, death stops you from accomplishing certain things.[5] The harm is not that you are deprived of time, the harm is that your autonomy is constrained by lack of time, so being deprived of those years is not part of the explanation of the harm. However, this is consistent with a deprivation account of the severity of death, for those years are an index of how badly you were harmed by dying at 9 rather than 90, even if we can't say you were deprived of those years (because you can't be deprived of anything when you don't exist).

The death burden argument

Now let's think about the death burden in a world where life extension is available. Does introducing life extension into a world where not everyone can get it increase the death burden for the Have-nots, who can't get life extension? Here is an argument that it does. If they had life extension, they might live for centuries. They cannot get life extension. Therefore, when they die, they miss out on centuries of life—a gigantic death burden. This would not be true if life extension were never developed. Therefore, introducing life extension into a world where not everyone can afford it makes the death burden immensely worse for the Have-nots. It increases the severity of their deaths. (Note that this argument works also for the kind of life extension that consists of slowing aging but not halting it, for the more we slow aging, the greater the life expectancy of those who get life extension and the greater the death burden for those who do not.) This isn't simply a matter of distress, for how a Have-not feels about a given harm is a separate, different harm. Rather, this may be part of what Have-nots are distressed *about*.

The impossibility objection

You might question the death burden argument. Instead of comparing death at 90 to how long a Have-not would have lived *with* life extension, why not compare it to how long he would have lived *without* life extension? After all, he couldn't get it, so why is it relevant? Right now it's medically impossible to provide life extension; in the future it will be fiscally impossible to provide it to everyone, at least for a while. Why does it matter whether you couldn't get it because it's not medically possible (the situation now) or because it's not fiscally possible (the situation in the future)?[6]

This objection rests on the fact that there is more than one counterfactual we might use to measure your death burden if you're a Have-not in a world where life extension exists:

A. how long you would have lived with life extension,

B. how long you would have lived without life extension, and

C. how long you would have lived with life extension, discounted by the odds of getting life extension (in other words, if you would have lived another 1,000 years but your odds of getting life extension were only 10 percent, then your loss is equivalent to 100 years).

The death burden argument uses counterfactual A (how long you would have lived with life extension), and the impossibility objection uses counterfactual B (how long you would have lived without it). Counterfactual C is a blend of the two. If A is the correct counterfactual, then introducing life extension into the world harms billions of people by increasing their death burdens, and if counterfactual B is the correct counterfactual, then introducing life extension into the world does not do this. (It may harm them in other ways, but right now, I'm focused solely on whether it harms Have-nots by increasing their death burdens.)

The argument for counterfactual A

The argument for counterfactual A is that we're really asking what *should* happen instead, not what's most *likely* to happen instead. For example, when we ask how long a patient would have lived, we might ask what would have happened in the cut-rate medical facility he could actually afford or we might ask what would have happened in the far superior medical facility that society should have paid for. Both counterfactuals are correct for different questions; the second counterfactual is correct for questions about justice in the distribution of healthcare. Similarly, counterfactual A is relevant in situations where we want to know what the world should be like—that is, what justice requires. The argument for counterfactual A is that we want to know whether introducing life extension into the world will be harmful to the Have-nots and hence unjust.

But this argument begs the question. Suppose we say that counterfactual A is the right counterfactual because the world should not be one where life extension is available only to the rich. Why should the world not be like this? Because making it available to some people but not to everyone harms those who don't get it, even if their lives are no shorter. Why does it harm them? Because that state of affairs is unjust; therefore, counterfactual A is correct. Why is that state of affairs unjust? Because it harms them, and around we go again. The justice argument for counterfactual A is circular unless we can find some reason other than the death burden for why unequal access to life extension is unjust.

Suppose it *is* unjust for some reason other than imposing a death burden on the Have-nots. Suppose, for example, that those who lack extended life have fewer opportunities and therefore failure to fund life extension for them violates their right to equal opportunity. Does that get us past the circularity? The idea is that if making life extension available is unjust to the Have-nots for some reason other than an increased death burden, then (and only then) counterfactual A becomes the correct measure for the death burden Have-nots suffer when life extension becomes available to the Haves. In effect, counterfactual A would piggy-back onto some other injustice.

I doubt that this works. Counterfactual A was introduced as a measure of how much death harms the Have-nots when life extension is available to the Haves, and it doesn't qualify as a measure of that harm just because, when we think about equal opportunity, we note that Have-nots would live far longer if they had life extension too. The harm that consists of being denied equal opportunity and the harm that consists of increased death burdens are two distinct and different harms, and establishing that one of those harms exists does not thereby establish that the other one exists any more than a criminal defendant acting alone becomes guilty of conspiracy to commit murder because he's also guilty of murder itself.

The argument for counterfactuals B and C

So the argument for counterfactual A seems faulty. What about counterfactual B? When we inquire into what might have been, we usually focus on what was most likely to happen: If you had taken that job, would you have made more money than you make now? No; that company went bankrupt in the recession, and you would have been forced back into the job market at a bad time. The argument for counterfactual B, then, is twofold. First, counterfactual B is the kind of counterfactual we usually take an interest in. When we want to know what might have been, it makes more sense to be interested in the alternatives that are most likely. Second, the argument for counterfactual A faced serious problems that counterfactual B does not face.

If counterfactual B is the right measure, then introducing life extension into a world where not everyone can get it doesn't increase the death burden for the Have-nots, for the Have-nots weren't likely to get life extension in that world. Because they weren't likely to get it, we can say, for example, that a Have-not who died at 85 didn't lose very many years, for she probably would have died by age 90 anyway.

However, counterfactual B suggests that life extension is simply impossible for the Have-nots, and that's not quite correct. Life extension would not be completely impossible for Have-nots to obtain. If you're a Have-not, you might win the lottery, or make a fortune in business, or manage to emigrate to a country rich enough to provide life extension for all its citizens, and thereby obtain life extension. There is a chance for you to get life extension; it's just a remote chance. For this reason, we should not use counterfactual B. We should use counterfactual C, where the death burden is measured by the years you would have had if you obtained life extension but discounted by the odds of your getting it. If, for example, your odds of getting life extension were 1 in 100 and life extension completely halts aging, then your death burden would be 1 percent of the estimated 1,000-year life expectancy for people who never age—a death burden equivalent to 10 lost years of life. You didn't lose that many years, but you lost something, and this is how to quantify it.

This needs some refinement. What are we calibrating the harm to? Your odds of getting life extension when society has done all it's morally required to do to make it possible for you to get it, or your odds of getting life extension when society has not done all it's required to do? It makes a difference. We don't want to say that the harm is less simply because you are unlikely to get life extension due to serious injustice in your society. That lets your society off the hook far too easily (the harm is one reason why this is unjust). The first answer, then: how hard it is for you to get it under conditions where your society has fulfilled its moral obligations to make it possible for you to get life extension. This does not mean we must assume your society can actually provide it, only that your society has done all it *can* to fulfill whatever obligation it has to make it possible for you to get it. That might include, among many other things, having a legal framework that provides equal opportunity to succeed in business so that you can make the money to afford life extension. For now, we can leave open the question of what obligations, if any, a society has in this direction. The point is that if we're going to use a counterfactual measure to decide whether people are harmed, we don't want to say there is no harm just because a society is unjust.

So the Have-nots do have a harm to be distressed about even before Malthusian consequences and other social ills arrive. However, the death burden is complex, subtle, and a good deal more abstract than most people are used to dealing with. How many Have-nots will have anything like this in mind when they express distress? They might say things like, "It sucks that the rich get this," "This is outrageous," or "I feel like my life means nothing now," but they're not likely to talk about counterfactual measures of the severity of death. In short, there *is* another harm here, but are Have-nots really aware of it? I'm inclined to say yes, at least for many of them. Many of them, I suspect, will feel as if their life spans are now less satisfactory, that their deaths are tragic in a way they were not before. They will feel that they're missing out on something that they wouldn't miss out on if life extension didn't exist. That, I think, is sufficient for us to conclude that their distress is about the death burden even if few of them ever see the arguments about the severity of death in the previous section.

7.4 Can we avoid making the death burden worse if we avoid developing life extension?

You may think that all this can be avoided if society simply refrains from developing life extension in the first place. You may even think that imposing larger death burdens on the Have-nots is a good reason to stop researching and developing life extension. We then face questions about whether there are moral arguments in favor of developing life extension that might outweigh that reason against it, but right now, I want to address a narrow question: Can the death burden be avoided by not developing life extension?

Unfortunately, this won't work—or at least not very well. The problem is that if life extension would eventually be developed if society did not prevent it, then life extension is possible even if society succeeds in preventing it. The possibility is more remote if society prevents it from being developed—to get it would require a lot of research and development and expense. However, it is still possible.

To see the problem from another angle, consider a range of cases:

Case 1 (a generation ago): Life extension is neither possible nor foreseeable, and you die at 90.

Case 2 (now): Life extension is foreseeable but not available to anyone, and you die at 90.

Case 3 (later this century): Life extension is available to some but you can't afford it, and you die at 90.

Case 4 (later this century): You have received life extension, and you die in an accident at 90.

Until very recently, everyone in human history lived in Case 1 (if they were lucky enough to die of old age). Though most of us don't realize it yet, we are beginning to find ourselves in Case 2, where life extension is foreseeable but not available to anyone. As the time when life extension is a practical possibility comes closer, the tragedy of missing out on life extension will become greater. It was not so bad to miss out on life extension a hundred years ago, but it's somewhat worse now, and it will be even worse as more and more potential life-extending drugs are in clinical trials.

This means that you—the reader—may already be suffering an increased death burden. In effect, you and I are temporal Have-nots (we were born too soon). We are living in a time when life extension is increasingly possible, and our death burden is increasing along with it. It is still heavily discounted, but as life extension becomes more possible, that burden will be discounted less and less. We are a sort of lost generation, for living closer to the time when life extension is developed makes our death burden worse than those of earlier generations, for whom life extension was a medical impossibility. Life extension is still a remote possibility for us, so our death burden is not yet very large, but it will grow larger the closer we get to developing life extension.

Even if society does nothing to develop life extension, it's still true that life extension is possible in our time, for it's possible for society to fund and pursue life extension research even if we don't do so. In other words, even if we do nothing, we may be lost. We can reduce our death burden by putting on the brakes and making life extension less possible, but we can't completely prevent our death burdens from increasing to some extent.

This means that those who get life extension in the future will do so at the cost of making death worse for others. In the long run, it's likely that a rising tide of prosperity and a falling cost of treatment will make life extension available to all who want it, but between now and then, humanity will contain some lost generations. The generations who live through the period when life extension slowly becomes a more and more imminent possibility will suffer, for their deaths will be worse than those of previous generations. Then again, even if we do nothing, it is still true that life extension is possible in our time, for it is possible for us to fund and pursue life extension research even if we do not do so.

Even if we do nothing, we are lost.

7.5 Conclusions

O. The distress suffered by Have-nots over the fact that others can get life extension and they can't is a harm, even if they have nothing to be distressed about. We may have a duty to alleviate it to some extent. However, it should be discounted unless there is another harm they are distressed about. Their distress is a separate harm from that other harm. (Section 7.2)
P. Making life extension available harms the Have-nots by increasing their death burden. The measure of the severity of death—of the death burden—is counterfactual C: How long you would have lived with life extension, discounted by the odds of getting life extension (in other words, if you would have lived another 1,000 years but your odds of getting life extension were only 10 percent, then your loss is equivalent to 100 years). (Section 7.3)
Q. Inhibiting or prohibiting the development and distribution of life extension can reduce the death burden, but it can never eliminate it. (Section 7.4)

8 The Have-nots—Equality and Access to Life Extension

8.1 Introduction

One of the most common objections to life extension is that the poor won't get it. The nightmare is easy to imagine: a world in which the wealthy few live on and on in endless youth, an elite gerontocracy of near-immortals who use their extra centuries to gather up the reins of property and power. They look down with indifference as generation after generation of short-lived ordinary people come and go in a passing spectacle of mortality, like ranchers watching a herd of cattle whose members change with the years. I call this the nightmare scenario.

Leonard Hayflick, famed for groundbreaking research in the biology of aging, sees this as the primary issue: "The first concern is that those involved in the discovery and the rich and powerful will have earliest access or, depending on availability, even the only access."[1] Many ethicists share his concern. Eric T. Juengst and his colleagues ask, "How should it be distributed? Serious ethical issues would arise if antiaging interventions were not universally available, but were distributed in response to status (economic, social, or political), merit, nationality, or other criteria."[2]

John Harris has raised this too: "The technology required to enable extended life-spans is likely to be expensive. Increased life expectancy would therefore be confined, at least initially, to a small minority of the population even in technologically advanced countries. Globally, the divide between high-income and low-income countries would increase."[3]

And Leon Kass: "Would it not be the ultimate injustice if only some people could afford a deathless existence, if the world were divided not only into rich and poor but into mortal and immortal?"[4]

No one disputes that this scenario is nightmarish, morally and otherwise. We will look briefly at why the nightmare scenario is unjust, but the most interesting questions concern what should be done about it:

• Should we make things equal (a) by subsidizing life extension for the Have-nots, thereby making sure *everyone* has access to it or (b) by making sure *no one* has access to it? Both approaches achieve equality; is one better than the other? (Section 8.3)

• Does justice require that we defer developing life extension until other needs are met first, such as the need for basic medical care, education, clean water, housing, and the like? (Section 8.4)

• Do the Will-nots have a moral duty to subsidize life extension for the Have-nots when they don't want life extension for themselves, don't benefit from having access to it, and may prefer to live in a world without it? (Section 8.5)

• What if the Haves will be able to subsidize life extension for some Have-nots but not all of them; is that any better than not subsidizing it for any of them? If not, what should they do instead? (Section 8.6)

• Suppose the Haves are able to subsidize life extension for some (or all) Have-nots and have a moral duty to do so, but we know that many of them will not do it. Does that justify trying to prevent the development of life extension, ensuring that the Haves will not fail in that duty by preventing the duty from arising in the first place? (Section 8.7)

8.2 Equality

The nightmare scenario is horrific, but why is it *unjust*? It is unjust because it's strikingly unequal in nearly everything a life can contain: opportunity, time, wealth, power, the amount of pleasure you can experience over time, the number and variety of desires you can satisfy over time, and so on. The Have-nots get less life experience, less travel, fewer careers, less opportunity for education, less time to work out their issues in psychotherapy, less time spent in meditation and spiritual pursuits, and less leisure time. They accumulate less wealth and less political power. The Have-nots will have less of many things. The poor already have a lot less, of course, but giving the Haves dramatically longer lives creates dramatically wider inequality. This seems obviously wrong to most people, and many people need no further explanation.[5]

Of course, inequality may not seem unjust to everyone, but those who have no problem with inequality—those who subscribe to extreme versions of libertarianism or overly simple versions of utilitarianism—may have no problem with the nightmare scenario either, and there's no point in arguing with them. If they do have a problem with it, then presumably they can use their own theories of justice to explain why it's wrong or stand accused of inconsistency.[6] That said, let's consider three egalitarian reasons why the nightmare scenario is unjust.

The first reason is that the nightmare scenario violates the Have-nots' right to equal opportunity. The more time you have, the more opportunity you have to acquire wealth, position, education, social relationships, and additional experiences and otherwise lead a fuller life. People with extended lives have a lot more time and therefore a lot more opportunity than others, and this is wrong.[7]

The second reason is that the existing distribution of wealth, power, opportunity, and so on is already unjust, and we should not make things worse. We should all agree on this, even if we disagree over precisely what the distribution should be. Letting the Haves live far longer than the Have-nots magnifies those unjust inequalities. There are two aspects to this.

For one thing, some Haves will use their extra time to consolidate their unjustly superior command of power and resources, partly by accumulating more wealth, partly by continuing to build their networks of political and commercial relationships, and partly by having more time to get better at doing these things. For another, the Haves will enjoy their unjust advantages over a longer period of time. If Bill Gates's children already have more money than they should, life extension enables them to enjoy and spend it over a longer span of time.

To see why this is worse, compare two situations. In the first, one person enjoys unjust superiority in income for 50 years, and in the second a succession of five people enjoy unjust superiority for 10 years each. Is the first situation worse than the second situation? In one way, no, for both situations involve 50 years of someone (or other) having more money than that person or persons should have. In another way, however, it is worse, for in the first situation, one person not only has more money; he has it for a much longer period of time and therefore enjoys an even wider margin of superior welfare over other people even if his welfare at any given time is not greater than the welfare the other five people enjoyed at any given time, one after the other. In this second way, life extension magnifies unjust inequality even if those who extend their lives have no more welfare per year than before.

The third reason the nightmare scenario is unjust is that although different theories of equality call for equality with regard to different things, all (or nearly all) of those theories imply that the nightmare scenario is unjust, for extended life gives the Haves more of the various things these theories tell us to distribute equally. The major contenders are equality of welfare theories, equality of resources theories, preference-satisfaction theories of equality, and capabilities approaches to equality. Extended life may not give you more welfare at any given time than you would otherwise have, but it does give you welfare over a longer span of time, and in that sense, it gives you more. Those who live extended lives can possess and use their resources over a longer span of time and also build on and augment their stock of resources by saving or investing them. Extended life gives people more time to benefit from satisfying their preferences and more time to satisfy more of those preferences. Finally, those who have extended life will possess and benefit from their capabilities over a longer span of time and have more time to develop, improve, and obtain capabilities. In short, time is a condition for all the things that various theories of equality say we should have equal amounts of; therefore, all these theories of equality are violated when some people get extended life and others do not.

Harry Frankfurt's *sufficientarianism* merits separate discussion because it raises a question about what counts as a reasonable expectation for one's life span. Sufficientarianism is not a theory of equality, for it doesn't require that everyone has the same amount of resources, but it does require that everyone has *enough* resources. This has an equalizing tendency by requiring us to fill the glass from the bottom up, so to speak.[8] Frankfurt defines *sufficient* as having enough that it is "reasonable for [a person] to be content."[9] And how much is that? It is a matter of that person's "attitude toward being provided with that much." Frankfurt defines this attitude as being satisfied with the amount of satisfaction you get from the resources you have.[10]

It's likely that some Have-nots will neither be content to die on schedule nor satisfied with their level of satisfaction about this. Some people who are *now* dying at a normal age are not happy about it, and this response will get worse, not better, once life extension is available. Once life extension becomes available in private clinics in Beverly Hills or the Cayman Islands, those who can't afford this may feel extreme dissatisfaction with their lot. The issue is not whether that dissatisfaction, and the distress it may give rise to, is a harm. The issue is whether that dissatisfaction is reasonable.

Is it reasonable? It may not seem reasonable now, but perhaps that's because we're now accustomed to a normal life span. Perhaps our sense of how much time is sufficient to live a full life will expand as lives themselves expand and as the Haves find new horizons to explore and new forms of maturity and achievement to reach for. We may someday decide that a decent life requires life extension just as we came to believe that everyone needs a telephone and motorized transportation. Once enough people come to see things that way, it may become reasonable to be discontented over lacking access to life extension, and a normal life span may no longer be sufficient. In short, whether sufficientarianism implies that the nightmare scenario is unjust turns on what is reasonable here, and that may be a matter of rising expectations.

8.3 Inequality as a reason for collective suttee

So the inequality in the nightmare scenario is unjust. What should we do about that? There are two ways to achieve equality where access to life extension is concerned: everyone gets access or no one gets access. In this section we'll discuss whether the requirement of equality is satisfied by denying life extension to the Haves so that no one gets it.

If we can't provide life extension to everyone, should we deny it to everyone? Several ethicists have suggested that developing life extension is a bad idea partly because not everyone will have access; this amounts to a suggestion that we avoid the nightmare scenario by not developing life extension, thereby denying it to everyone. Walter Glannon is explicit about this:

Of course, we could allow people to pay for life-extending technology. But this would be unfair to those who could not afford to pay because it would preclude them from having the same choice as those who are financially better off. [This provides some] moral grounds for not allowing people to extend their lives beyond the present norm.[11]

I call this *collective suttee*: just as widows in traditional India were not allowed to outlive their husbands (*suttee* involves putting them on the husband's funeral pyre and burning them to death), so potential Methuse-lahs must not outlive the rest of us. Collective suttee consists of prohibiting life extension on the grounds that it's wrong for some to have access when it's not possible to give access to everyone. Collective suttee is a version of what's often called "leveling-down": achieving equality by bringing every-one down to the same level as the worst-off people without making anyone better off. It is leveling-down if we prohibit life extension once it has been developed, and it is leveling-down if we prevent the development of life extension in the first place, provided the intent is to ensure that inequality does not occur and this happens in a way that makes no one better off. Both a policy of inhibiting the development of life extension (Inhibition) and a policy of preventing life extension from being used (Prohibition), if done out of concern for equality, are forms of leveling-down.

Does egalitarian justice require leveling-down even when that makes no one better off? One common objection to egalitarianism is that it does, at least sometimes. This is an objection because leveling-down is considered obviously wrong. If equality requires something that's obviously wrong, then egalitarianism itself must be wrong. Therefore, either egalitarianism is false or egalitarianism must be modified so that it doesn't require leveling-down. Those who raise this objection consider leveling-down so obviously wrong that we don't need to explain why it's wrong, so they don't say much about why leveling-down is wrong.

Some philosophers with egalitarian leanings respond by adopting a nonegalitarian theory that requires equality in most situations but doesn't require leveling-down, such as sufficientarianism (giving everyone enough, even if there is still inequality) or prioritarianism (giving priority to those who are most in need). Other egalitarians respond by arguing that even if egalitarianism requires leveling-down, there are other moral principles that prohibit leveling-down and those principles sometimes outweigh the value of equality, so a complete moral theory will not call for leveling-down. Joseph Raz, for example, says that leveling-down is wrong because it's wasteful, and wasting resources is wrong.[12] In other words (according to this response), it's unfair to pin this problem on egalitarianism. It's handled by another part of morality.

But our concern is not how to defend egalitarianism against the leveling-down objection. The point is simply that no one thinks equality should be achieved by leveling-down. I will assume, then, that the reader agrees that leveling-down is the wrong response to inequality. It follows that there is no reason to think that justice is served by denying life extension to the Haves. Therefore, collective suttee is a bad idea. Justice does not require Inhibition or Prohibition if the point of those policies is to prevent the nightmare scenario by making sure no one gets life extension.

This conclusion is also supported by the fact that it fits our judgments about other issues in healthcare. For example, there aren't enough hearts available for transplant for everyone who needs one. We can't achieve equality in that respect by providing more hearts; the only way to achieve it is to ban heart transplants. We do not, however, deny heart transplants to everyone on the grounds that there aren't enough hearts for everyone who needs one. Similarly, we should neither prohibit life extension nor inhibit its development just because it won't be possible to provide it to everyone who wants it.

8.4 What if other needs are more pressing?

Sometimes people argue that we should not devote resources to developing life extension until we have first met other needs. The argument is that the other needs are more pressing: We should not spend money and resources pursuing life extension when there are so many more important needs out there. Around the world, there are hundreds of millions who lack food, clean water, and basic medical care. Here at home, many people lack sufficient education, money for retirement, meaningful employment, and so on. Our roads and infrastructure need repair, and our environment needs to be protected and brought back from a damaged state. We should devote resources to bringing the poor up to middle-class standards before we let the rich live extended lives and fight diseases that shorten life before extending lives beyond a normal span. I call this the *pressing needs objection*.

Among other things, this objection tells us that we should not consider justice and life extension in isolation. We must think about the distribution of life extension in the context of distributive justice for all other goods too. Failure to do this is like thinking about justice in the distribution of real estate while ignoring justice in the distribution of money, political power, opportunity, and many other things. To that extent, I agree with the argument behind this objection.

However, it does not follow that life extension should not be made available until other, more pressing needs have been met. The reason is that the pressing needs objection takes the point of view of what I will call a *total planner*, someone who has the power to arrange the distribution of all goods. Let me invoke Rawls's distinction between ideal theory and nonideal theory. Ideal theory concerns justice in societies where citizens and the society at large are able and willing to comply with principles of justice, while nonideal theory concerns justice in societies where this is not the case.[13] What I am calling a "total planner" is working in the kind of society ideal theory is about. A total planner who has control over the distribution of all goods to all people may well have good reason to give priority to goods other than life extension. However, even if that's the right thing for a total planner to do, those who can influence the distribution of life extension are not in the position of a total planner. The social decisions about whether to develop life extension and how to make it available will be made by what I will call *partial planners*, such as private parties, politicians, and government officials. Such people have only limited control over their society; they work under nonideal conditions. Privately, they may believe that their social order is unjust in many ways, but they aren't able to impose what they believe is the most just total plan on everyone else.

Philosophers working on theory of distributive justice tend to take the perspective of total planners (they are working on a total theory, after all), but philosophers working in applied ethics must ask what justice requires in the real world, where societies are partially unjust. That requires the perspective of a partial planner, and partial planners should make different choices than total planners. It may be true that total planners should allocate social resources to other needs first and that there will not be enough left over to fund life extension research, development, and distribution. However, we are not total planners. We must seek justice in a partially unjust world.

What does justice require of partial planners in a partially unjust society? It does not require that they prioritize the distribution of goods in ways that are appropriate only for total planners. To see why, let us make two assumptions. First, assume there are goods (besides life extension) that are distributed in a way that's unjust. Second, assume that life extension will be distributed in a way that *is* just. I'm not making the second assumption because we can safely assume this will happen. I'm making it because I want to know whether the fact that other goods are distributed unjustly means that distributing life extension justly is not enough. By a "just distribution" of life extension, I mean a distribution that *would* be just in a society where all other distributions are just. Our question is whether a just distribution of that kind becomes unjust when other distributions are not just.

Why might injustice in the distribution of other goods make it unjust to make life extension available, even if the distribution of life extension itself is otherwise just? The reason can't be that denying life extension to everyone makes it more likely that other, more pressing needs will be met. That's unlikely to happen, for any society with the discipline to withhold life extension until other injustices are corrected also has the discipline to go ahead and correct those other injustices instead of waiting for them to be corrected by someone else. This isn't a problem from the perspective of a total planner, who has control of all resources, but it's a serious problem from the perspective of a partial planner, who does not. The partial planner must ask whether her decision to inhibit or prohibit life extension will make it more likely that other goods will be distributed more justly.

The answer is almost certainly no. Yes, it's possible that delaying or preventing life extension from being developed or preventing the rich from getting it will put pressure on enough wealthy and powerful people to finally address other injustices. However, it is much more likely that such people will use their wealth and power to resist such efforts and still get life extension. After all, that's their track record. Moreover, if holding back some good does not make it more likely that other, more pressing needs will be met, we are not likely to see any other reason to hold it back. We don't ban heart transplants just because the poor lack basic medical care.

In addition, this argument does not generalize well. Assume for discussion that if one good is unjustly distributed, we must ban or prohibit the distribution of other goods even if those other goods are distributed in a way that is just. If the fact that one good is unjustly distributed requires us to ban a second good even when that second good will be distributed justly, then *no* goods can ever be made available at all, even when distributed justly, unless *all* goods are distributed justly. In short, this means that it's better not to distribute any goods at all rather than distribute some goods justly and other goods unjustly. That, of course, is ridiculous, and a claim that implies a ridiculous conclusion is probably false. Partial justice is preferable to denying all goods to everyone. Therefore, the fact that other goods are not distributed justly is not a reason to deny life extension to the Haves.[14]

8.5 Who has a duty to subsidize life extension for Have-nots?

Who has a duty to do something about unequal access to life extension? We could say "society," but society is composed of Haves, Will-nots, and Have-nots. I take the position that if justice says that you should have less than you actually have and society has not addressed that injustice, then you're not off the hook simply because your society is not fully just. You have a moral duty to make the redistribution yourself, so far as your own share of things is concerned. Moreover, even if society does make the distribution that justice requires, you still have that duty—it's just that society has performed it for you or forced you to perform it through your taxes or some other mechanism. We could speak of justice in the abstract here and talk about which groups justice requires we take from, but I prefer to speak of a *duty of justice*, for that's a less clumsy way to talk about which groups justice requires taking something from. To be more precise, I will speak of a *duty of equality*—a duty to alleviate or prevent unjust inequality. Let's assume, to keep things simple, that those who are not well-off have no duty to subsidize life extension for the Have-nots and that the Have-nots include all and only those who are not well-off. This means that the only members of society who have a duty to pay such a subsidy are Haves, Will-nots, or both.

Before we decide who has the duty, let's start by considering what this duty requires. Suppose you have more wealth than you would have if your society had a just distribution of wealth. You then have a duty to give away that extra wealth so that you no longer have more than your fair share. In effect, you have no right to that portion of your wealth, so you should restore it to those who have a right to it. Failure to do so is tantamount to theft. However, this duty does not require you to do *everything* you can to make society as equal as it should be. Suppose, for example, that you are a wealthy person in a society where a just distribution of wealth would leave you with half as much wealth as you now have. You have a personal duty of equality to give away half your wealth. Presumably this would leave you worse off than before but still better off than most of those who have too little, for others are not giving away any of their wealth and many people still have less than they should have. You have a duty to alleviate unjust inequality up to the point where you have what a just distribution would give you, but not a duty to give away more than that, even though giving away more than that would bring society even closer to what a just distribution requires. If your neighbor still has less than she should, you don't have to adjust your position below what justice requires in order to bring her closer to what she should have (by, for example, transferring enough of your wealth so that you both have 75 percent of what a just distribution would require for both of you). The reason, of course, is that justice can't require people to have less than what justice requires them to have.

The Haves

Now consider the Haves. When they get access to life extension, they don't immediately get more resources than they had before; they simply have an opportunity to use those resources to buy something they could not get before. They are better off in a way, but does access to life extension make them better off in a way that requires them to do more for the Have-nots than justice required before life extension was developed? It's not as if they each had X amount of wealth before life extension was developed and now they have X + n amount after it's developed. Rather, they still have X amount of wealth, but they can now spend that wealth on life extension instead of spending it on other things (including real estate, fancy cars, business investments). If they have no more wealth than before, why does their duty of equality demand a larger transfer of wealth to the Have-nots than it did before life extension came along?

One possible answer is that it does not. They already have a duty to transfer some of their wealth (a duty society might enforce through taxation), and now that life extension is possible, society should use some of that wealth to subsidize life extension for the Have-nots. The total transfer from Haves to Have-nots would be no greater, but it would be spent differently.

However, there are reasons to think the advent of life extension means the duty of equality requires even more from the Haves—a larger transfer—than it did before. First, the Haves *are* now wealthier, at least in a nonmonetary sense. They now have more time to enjoy their wealth, use their wealth, and enjoy whatever life provides. Even if their monetary wealth doesn't increase at all, enjoying their wealth (and everything else life contains) over a longer span of time is effectively an increase in their wealth, or in what that wealth can buy for them. Second, they have more time to accumulate yet more wealth. Third, even if getting life extension doesn't give the Haves more of whatever goods justice is concerned with, it does widen the gap between Haves and Have-nots. In other words, even if a Have does not have more goods as a result of extending her life, the Have-not living next door is nonetheless worse off compared to the Have, even if that Have-not is not worse off in a noncomparative sense. Society is now more unequal. If equality requires transfers of goods to make things more equal, then when life extension makes things even more unequal, justice requires larger transfers of wealth and other goods to reduce that increased inequality.

The Will-nots

So making life extension available increases what the duty of equality requires from Haves (and even if it doesn't, whatever should be taken from them and redistributed should still be spent partly on life extension subsidies for Have-nots). But what about the Will-nots? They too have enough resources to use life extension, and they're just as well-off as the Haves (at least starting out, before the Haves get all that extra time), for economically they're the same group. Perhaps they have a duty to use their privileged position to make things more equal for the Have-nots, since they have the same capacity to do so as the Haves—perhaps even more so, since they are not spending any of their resources on life extension.

Yet it seems odd to require Will-nots to subsidize life extension for others when they don't want it for themselves and may even prefer to live in a world where there is no life extension at all. Now they must not only live in that world but also help compensate those who lack access to the very technology they disdain. How can this be just?

As with the Haves, there are two ways someone might claim that the advent of life extension affects the Will-nots duty of equality. First, one might claim that the advent of life extension requires reallocating whatever transfers of wealth they must already provide. In other words, if justice requires transferring 20 percent of some Will-not's income, that money must now be spent partly on life extension (taking funds away from the other things that tax money has been spent on). In that case, the advent of life extension doesn't increase what justice demands from the Will-nots, though some of them may still protest having to subsidize something they don't approve of.

Second, someone might claim that the advent of life extension increases what the duty of equality requires from Will-nots. This claim is not correct. Why might someone think it is? The reason can't be that the availability of life extension makes Will-nots better off; they don't want it and turn it down. The reason must be this: The advent of life extension widens the inequality between Have-nots who want life extension and the Haves who extend their own lives.

But the Will-nots neither benefitted from that wider gap in equality nor brought it about. The wider gap in equality caused by developing life extension occurred without their effort, intent, or consent. (This is a bit too broad; some of them may favor developing it even though they don't want it for themselves, but let's focus on those who don't favor developing it.) The fact that they neither caused that gap nor benefitted from what caused it (because they decline to extend their lives) are reasons to conclude that their duty of equality does not become more demanding simply because that situation has occurred.

Moreover, the gap is not wider between the Have-nots and Will-nots; it's wider between the Have-nots and the Haves. If that wider gap increases what the duty of equality demands, that increased demand should fall on those who are better off relative to the Have-nots (the Haves, with their longer life spans) and not on those who are *not* better off relative to the Have-nots than they were before (the Will-nots, whose life spans are similar to those of Have-nots). It's as if an economy evolved in a way that made the top third twice as rich as before; the middle third would not thereby have a greater duty to help the bottom third.

Thus, the duty of equality to adjust for unequal access to life extension falls solely on the Haves to the extent that duty becomes more demanding in response to the advent of life extension. It falls on both Haves and Will-nots to the extent that justice already requires some level of transfer from the better off to the less well off, and justice requires that society reallocate some of that money to provide life extension for Have-nots. However, the size of the transfer justice requires will not be larger for the Will-nots simply because the Haves now have life extension. It will be larger only for the Haves.

This has implications for how to fund access to life extension for Have-nots (or how to fund whatever compensation they might receive in place of life extension). Society should not tax the Will-nots any more than what justice requires in that society before the advent of life extension; they have no duty to remedy this particular inequality. Society may use those taxes to provide life extension to Have-nots, but it may not increase those taxes to do so (unless the taxes are currently lower than what justice otherwise requires). Therefore, if that society is already transferring as much as justice requires before the advent of life extension, taxes should not be increased but spending may be reallocated.

The Haves, however, must contribute more. One way to do this is to impose a use tax or user fee on life extension treatments or dramatically increase the income tax for anyone above an age that no one can reach without life extension. That income tax could even go up the longer you live, a kind of temporally progressive tax. These extra revenues received from the Haves can be used to subsidize life extension (or other compensation) for Have-nots. This could be set up as a sliding scale, with heavier fees on those who are better off, smaller fees on the less well-off, and a scale of subsidies for the Have-nots according to need.

8.6 What if it's possible to provide access to some Have-nots but not possible to provide it to all of them?

It may turn out that the Haves can't afford to provide life extension to *all* Have-nots, that there is simply not enough money in the world to do that. However, this doesn't mean that they don't have to do anything at all for the Have-nots. They could, for example, subsidize life extension for *some* Have-nots even if they can't subsidize it for all of them. This would ameliorate the nightmare scenario to some extent, like an Ivy League university setting aside slots for disadvantaged students in order to atone for social disparities. In that case, there will still be Have-nots (whoever didn't get the subsidy) but fewer of them. In section 8.3, I argued that it's permissible to have a world where the Haves get life extension and the Have-nots don't so long as it's not possible to give it to them. If, however, the Haves are able to compensate some Have-nots but can't reasonably afford to compensate all of them (either because justice doesn't require that large a transfer or because there simply isn't enough money in the world to do so), then they must do precisely that. They must do as much as they can, even if they can't provide life extension to everyone.

What does justice require when it comes to choosing which Have-nots to benefit when there is no duty to benefit all of them? Should it be done by lottery? Should societies start with their own citizens so that rich countries subsidize their own Have-nots before extending help to Have-nots in poor countries? Should the criteria take into account whether a Have-not is partially responsible for not being a Have? Should we start with the Have-nots who are worst-off overall and work our way up from there? I'll set these questions aside, for there's a better way to approach this.

It's more equitable to benefit the Have-nots in some way that does not require helping some and not others. Because it's more equitable, this is what the duty of equality requires, even if that means that no Have-nots will get life extension subsidies instead of some getting subsidies and some not. Haves can reduce inequality for Have-nots in many ways besides giving them life extension, such as paying compensation, providing extra social services, and so on. We might, for example, impose a dramatically higher income tax on anyone who has received life extension treatments; as soon as you start the treatment, your taxes triple, and 90 percent of your estate is confiscated for redistribution (you have plenty of time to accumulate another one). (Presumably the Haves would be left with enough wealth to continue receiving life extension, not because justice requires letting them have it, but because this scenario arises only if justice does not require taking so much away that they cannot buy life extension for themselves in the first place.) That extra tax revenue could be spent on any number of things the Have-nots need besides life extension. The goods it pays for can be divided equally so that all Have-nots receive roughly equal benefits. If the Haves and Will-nots do not have a duty to make life extension available to everyone because that's not possible, then they have a duty to make the Have-nots less unequally situated by equalizing the distribution of things besides access to life extension, things that can be provided equally to all Have-nots.

Is this a kind of leveling-down, so far as the lucky Have-nots (the ones who would otherwise have received subsidized life extension under some redistribution) are concerned? Not really, for there is no determinate group of Have-nots who possess access and then find it taken away from them. Leveling-down can happen only to those who have more, and before a distribution is made, there *are* no Have-nots who have more. If this were not true, then we could justify an inequality simply by creating it and then claiming that eliminating it would be leveling-down and hence impermissible.

Therefore, justice does not require compensating the Have-nots by providing life extension to some of them but not others. If we cannot provide life extension to all the Have-nots, then justice requires compensating them with something other than life extension—something that can be distributed equally among them.

8.7 If we are sure that many Haves will breach their duty to the Have-nots, is that a reason to deny it to everyone?

What if we can be pretty sure that many or even most Haves will not perform their duty of equality to make things more equal for Have-nots in whatever way is possible, not because they can't but because they simply fail to do so? ("Pretty sure" is putting it mildly. I think we can count on it.) Does *that* justify Inhibition and/or Prohibition? There are five reasons why the answer is no.

First, this cure for breaches of duty is worse than the disease, so to speak. Yes, preventing life extension from being developed does prevent unequal access to life extension and thereby prevents breaches of the duty to provide equal access. However, preventing life extension from being developed prevents breaches of the duty to subsidize life extension for Have-nots in a purely technical sense—not by ensuring performance of that duty but by eliminating the conditions for the duty to arise at all. This is like outlawing medicine so that the rich will not fail to subsidize healthcare for the poor.

Second, preventing inequality when doing so doesn't make anyone better off is a kind of preventive leveling-down. Leveling-down is not required by the duty of equality; therefore, leveling-down is not required to prevent inequality either.

Third, some Haves are generous and will subsidize life extension for the Have-nots. Inhibiting life extension prevents breaches but also prevents performances of the duty of equality. Inhibition is unfair to the generous Haves.

Fourth, we can't justify this as punishment for stingy Haves, for we can't justifiably punish someone for something they have not yet done.

Moreover, preventing life extension from being developed also punishes generous Haves and the Have-nots those generous Haves would otherwise help, and they should not be punished either.

Fifth, inhibiting the development of life extension delays the day when life extension may be cheap enough and the world's economy prosperous enough to provide life extension to everyone, thereby hurting people in certain generations who could have had life extension if it had been developed sooner. Almost all new technologies start out expensive and later come down in price, often quite dramatically, especially if the market is very large and we find ways to automate the production. This has happened with cell phones, pocket calculators, and some pharmaceutical drugs.[15] There is reason to anticipate that this will happen with life extension too. If we let life extension research and development move forward as quickly as it can, the day when everyone can receive life extension will arrive that much sooner.

8.8 Conclusions

R. A world where the Haves get life extension and the Have-nots don't is unequal in ways that are unjust. (Section 8.2)

S. Once life extension has been available to Haves for long enough, it may become reasonable for people to think that extended life is, so to speak, normal—that having enough in life includes having extended life and that a normal life span is no longer sufficient. (Section 8.2)

T. Equality doesn't require leveling-down, so the fact that life extension will not be available to everyone is not a justification for retarding the development of life extension (Inhibition) or preventing people from having it (Prohibition). (Section 8.3)

U. We should not postpone or prohibit life extension until other, more pressing needs are met first. For example, the fact that some people lack basic healthcare is not a good reason to prevent life extension from being developed or distributed. (Section 8.4)

V. Even if society should redistribute resources from wealthier members to poorer members, the advent of life extension does not increase the size of transfer that justice requires from Will-nots, though it may require reallocating that transfer. However, it does require increasing the size of transfer required from the Haves, perhaps by imposing a user fee or increased tax on those who extend their lives. (Section 8.5)

W. If it's not possible to provide life extension to all Have-nots who want it, justice is not served by providing it to some Have-nots but not others. Instead, justice requires that Have-nots should be compensated in some way that can be distributed equally among them, such as a redistribution of wealth or increased social spending that benefits all of them. (Section 8.6)

X. Even if we can be sure that many Haves will evade their duty to subsidize or otherwise compensate Have-nots, Prohibition and Inhibition are not justified. (Section 8.7)

9 Deciding among the Groups—Maximizing Welfare

9.1 Introduction

There are two levels to life extension ethics. In chapters 2 through 8, we stayed entirely on the first level. In this chapter, we move to the second. We will stay on the second level from here all the way through chapter 11.

On the first level, we asked how each of our three groups would be affected if life extension becomes available to the Haves. We discussed whether extended life is good for the Haves, whether making life extension available to the Haves harms the Have-nots through distress and increased death burdens, and whether it harms the Will-nots through moral injury and reduced death benefits. We also considered potential good and bad consequences for everyone, particularly Malthusian consequences. Finally, we considered whether it's unjust to make life extension available to some people when not everyone can afford it and whether justice requires making life extension available to those who can't afford it. I argued for certain conclusions on each of these issues.

These conclusions pull in different directions. Some of the conclusions favor Promotion (funding life extension research aggressively) and have proextension implications in general. Some of them favor Inhibition (inhibiting the development of life extension as much as possible) and have antiextension implications in general. Some of the conclusions don't clearly favor one policy or the other. In general, I concluded that developing life extension and making it available is, on balance, good for the Haves, bad for the Have-nots, and may impose small harms on some of the Will-nots.

So we have reasons in favor of developing life extension and reasons against doing so. However, we can't leave it there. We need to settle on a single policy for funding life extension research, and that requires a single overall conclusion about whether a world with life extension is, all things considered, better than a world without life extension. Reaching that conclusion happens on what I call the "second level" in life extension ethics, where we decide whose interests have priority and why.

How do we choose between Inhibition and Promotion—or, more generally, between a world with life extension and a world without it? We could put the issue up for a vote and settle it democratically. However, we want to know what's morally right, and what the majority votes for is not always morally best. Moreover, voting provides the policy the majority wants, but that might be the wrong policy if it violates the rights of a minority. Finally, even if this were decided by vote, many people would want to know what justice requires before they decide how to vote. This is not to say that this choice should never be voted on, only that we can't tell what morality requires by seeing how the vote turns out.

It's likely that the Have-nots and Will-nots will outnumber the Haves for some time (until the world grows richer and life extension gets cheaper), but we can't just count the number of people and select the policy that benefits the majority. Sometimes morality requires favoring the interests of a smaller group of people over the interests of a larger group. For example, we would not force a minority to work for free in order to increase incomes for the majority.

So we can't just put this up for a vote or count people. We have to make some judgments about the moral weight of the interests and welfare of all three groups and decide whether to institute a policy of Promotion (funding life extension research aggressively), which favors the Haves over the other two groups, or a policy of Inhibition (underfunding life extension research and otherwise trying to prevent it), which favors the other two groups over the Haves.

We'll start by asking which world has greater total welfare: a world with life extension or a world without it. Before we get to that, however, we must look at the relationship between welfare and the rest of moral theory.

9.2 Midlevel principles, moral theory, and doing applied ethics

One obvious way to adjudicate among such conflicting interests is to identify the correct ethical theory and apply it. However, I won't try to do that. Instead, I'll appeal to a few midlevel moral principles and some rights. This set of principles and rights should be acceptable from the standpoint of most major ethical theories, though various theories might state them in very different ways.

The first midlevel principle is that we should try to maximize welfare. (I will sharpen this claim later.) This is similar to the principle of utility, but without the implication that maximizing utility is the whole of morality. We might call this a principle of beneficence; most moral theories include something like it. Kant, for example, used the categorical imperative to argue that we have a duty of beneficence (though the reasons why we have a duty of beneficence are grounded not in beneficence but in Kant's conception of a free will and the moral law). Ross included a duty of beneficence among his prima facie duties, and Hume argued that beneficence is among the natural virtues.

However, I will also appeal to rights, which limit when and how we can maximize welfare. For the most part, we must maximize welfare without violating any rights. In other words, in most situations where we have a choice between (a) producing a particular amount of welfare without violating any rights and (b) producing some larger sum of welfare by violating some right, we should do (a), respecting that right and producing the smaller sum of welfare. As I'll argue later on, there is an exception to this: if the amount of welfare gained by violating a right is very, very large compared to the importance of whatever interest the right protects, then maximizing welfare may be the right thing to do even if there is no way to avoid violating that right.

I will discuss three rights. Each of these rights has a corresponding duty. First, we have a *right against harm* and a corresponding duty not to harm others (a duty of nonmaleficence). Second, we have a right to determine our own lives, a *right of self-determination*, at least when it comes to things that primarily affect ourselves and affect other people only secondarily or indirectly and provided that we don't harm others. There is a corresponding duty not to interfere with someone else's self-determination. (This duty is sometimes expressed in terms of a principle of respect for autonomy.) This duty is limited, however; we may not interfere with efforts at self-determination, but in most situations, we have no duty to assist those efforts. Third, we have a *right to equality* (a right to equal treatment, if you prefer) and a corresponding duty of equality. We can think of these three duties as three more midlevel principles.

Instead of trying to identify *the* correct ethical theory and then apply it, I will apply this set of four midlevel principles. This set of principles is not meant to be an ethical theory. Instead, it's meant to be consistent with most ethical theories. It's also meant to enable us to stay out of controversies over which ethical theory is correct. To see how it does this, we must look more closely at what ethical theories are and how this set of principles is related to them.

Ethical theory (in the sense I have in mind) is the project of identifying the common element or elements in our correct moral judgments at the highest level of abstraction, a level beyond which no further reduction to some simpler or smaller element is possible. (Some philosophers define "ethical theory" more broadly, in a way that may include some midlevel principles. I am using the term more narrowly.) Such theories include Kantian theory, several varieties of utilitarianism, various kinds of virtue ethics, Rossian intuitionism, and many others. Kant, for example, argues that the categorical imperative is the fundamental criterion of morality, while Mill says it's the principle of utility. Ross did not reduce all of morality to a single principle or duty; he reduced it to nine duties. Various virtue theorists have reduced it to a small set of virtues. The set of principles I am using would be an ethical theory if there were no more fundamental principles or criteria they share or derive from or to which they could be reduced. However, I am not committed to thinking of them that way, and I leave open the possibility that they might be reduced to or derived from something more fundamental (like the principle of utility, the categorical imperative, or something else).

Ethical theories are evaluated for consistency with our considered moral judgments about clear cases and with firmly held moral principles, such as "do not harm" and "do not steal." Any ethical theory that tells us it's permissible to harm people without some overriding moral reason that outweighs the harm, or that stealing is generally okay, or that collides with a high percentage of our more confident moral judgments about cases has some explaining to do.

The principles I will be using are among the firmly held moral principles that are used to evaluate the plausibility of ethical theories. Any ethical theory should be consistent with and imply claims that welfare is good, that having more welfare is better, that we should not harm, and that self-determination and equality are important. Moreover, the fundamental element should explain and imply these principles as consequences of that element: for example, they might be utility-maximizing if we deploy them in moral practice, or they might be versions of Ross's duties, or they might be guidelines for conduct in accord with the virtues. Finally, these principles should imply the verdicts in clear-cut moral cases that are both consistent with our considered moral judgments about those cases and consistent with what the fundamental element tells us about them.

So these midlevel principles are supposed to be neutral concerning which ethical theory is correct. That said, there is one exception to that neutrality. Although most moral theories include something like these rights, utilitarianism does not. However, utilitarianism may include moral rules of thumb used in a heuristic fashion and which, at the level of moral practice, look a lot like rights. Mill and most other utilitarians believe that we use various rules in moral deliberation partly because we are prone to bias in our favor and partly because we often lack the time, ability, or information to accurately estimate net utility. Rule utilitarianism is a version of this view. Such rules are ultimately attempts to maximize utility. If we follow them, we are more likely to maximize welfare than we would be if we tried to estimate net utility directly, without using such rules. For example, Mill's *On Liberty* is an extended utilitarian argument for treating liberty as something like a deontic constraint, even though Mill expressly rejects rights.

However, even a utilitarian who advocates using rules in most situations (rather than trying to estimate which act will maximize net utility) could hold that there are times when we should set rules aside and think about net utility directly: when considering major questions of public policy, where we do have the time and resources to gather information and take our time in deliberation. *That* is a version of utilitarianism (or at least an aspect of it) my set of midlevel principles is not consistent with. I reject that approach, for I believe that using welfare-maximization alone, without any rights or other deontic constraints, sometimes produces results that, I believe, conflict with the considered moral judgments most of us have about equality, harm, or self-determination. Aside from versions of utilitarianism that use act-utilitarianism for large public policy choices, these midlevel principles enable us to avoid taking sides in controversies over which ethical theory is correct.

Now we're ready to apply the four midlevel principles and decide how to resolve conflicts among the interests of the three groups. There are three stages to this. The first stage is to ask which policy—Inhibition or Promotion—maximizes welfare. We will tackle that in this chapter. The second stage is to identify the relevant rights of the Haves, the Have-nots, and the Will-nots. We will do that in chapter 10. The third stage is to ask whether the welfare Promotion will produce outweighs the rights it conflicts with. We will do that in chapter 11.

9.3 What it means to maximize welfare

This chapter is focused on welfare. All moral theories have something like a concept of welfare. *Welfare* is a generic term I use to indicate whatever constitutes well-being for humans. Utilitarians call this "utility." There are many definitions of welfare (utility). Some define it as pleasure, or happiness. Some define it as preference-satisfaction—getting what you prefer to have, whether or not that makes you feel happy. Some define it as getting things that are objectively good for you (such as education, health, good social relations, and many other things), regardless of whether you enjoy those things or want them. I'm going to assume that getting more years of healthy life gives you more of all those things, so for our purposes, we don't have to choose among those three ways of defining welfare.

Welfare is important in morality, and all else being equal, the more of it, the better. Our moral theory requires us to ask (among other things) which policy—Promotion or Inhibition—will produce a greater amount of welfare. There are two broad ways to understand what it means to maximize welfare. (I'll borrow from utilitarianism for these accounts. We can use a utilitarian account of welfare while rejecting the utilitarian tendency to see maximizing welfare as the core of morality.)

According to *total utilitarianism*, all else being equal, a state of affairs is better the greater the net sum of welfare it contains, and the best state of affairs has the largest net amount of welfare. To measure the welfare produced by some action or policy, add up the net welfare of each affected individual. According to *average utilitarianism*, a state of affairs is better, all else being equal, the higher the average level of net welfare per person (I'll leave out complications involving beings who are not persons), and the best state of affairs has the highest average net welfare per person. For example, if one state of affairs has 1,000,000 people and an average welfare level of N and another state of affairs has 100,000 people and an average welfare level of 2N, then the second state of affairs is morally better even though the total welfare in the second state of affairs is smaller. (If both populations are the same size, or the one with the higher average is also larger, then total utilitarianism and average utilitarianism have the same implications; they differ only when the population with the higher average is smaller than the one with the larger total.) Each of these views faces serious objections, but I won't take sides. It turns out that both accounts of maximizing welfare imply the same conclusions about the most likely life extension scenarios, so we need not choose between them here.

Because we're talking about lives of different lengths, we also need to think about how to estimate the total welfare per life. I will use the number of years lived as a rough index of the total welfare a person receives over the course of their life. This is, of course, an average; some years contain a lot more welfare than others. It's also true that some people enjoy higher welfare per year than other people, but we'll average that out over a population. I'll discuss the amount of welfare in units of *life-years*. A person who lives, for example, 85 years, will be assumed to have 85 life-years' worth of total welfare in her life; a person who lives twice as long will be assumed to have 170 life-years of total welfare over the course of her life, and so on.[1]

There is a possibility, of course, that each additional year of life will contain less welfare over time, if that person lives for decades or centuries beyond a normal life span, perhaps because she couldn't avoid chronic boredom. However, it's also possible that people who live for decades or centuries longer will learn how to live more fully in the present, to appreciate life better, to understand what is truly of value and what is not, and thereby enjoy more welfare per year than they did in their first century of life. We really can't say whether the average welfare per life-year will decline over time for an entire population. Estimates of welfare across a population could be off in any number of ways. The percentage of Haves might be more or less than I suppose, and their life spans might be more or less than I suppose. For these reasons, I'm going to ignore that complication and assume that life-years do not contain less welfare per year in an extended life.

9.4 Objection: we don't have enough information

In section 9.5, I'll try to estimate the amount of welfare that Promotion and Inhibition are each likely to produce, and I'll throw out some surprisingly exact figures. I'll assume that in the near term, 10 percent of the world's population will be able to afford life extension and 90 percent will not. I'll assume that each Have-not's welfare is cut in half if 10 percent of the population lives extended lives. I'll argue that if only 5 percent of the population had a life expectancy of 1,000 years, the net welfare would be equivalent to 9,037.5 life-years of welfare for every 100 people (5 Haves, 95 Have-nots). I'll make a lot of other, similarly precise-sounding claims. This raises an objection I want to deal with before we go any further.

I realize that the precision of these numbers is ludicrous. We have nowhere near enough information about these matters to know with anything approaching confidence what the actual percentage of Haves, Have-nots, and Will-nots in a population will be, or what life extension will cost, or how much each group suffers per year compared to the other two groups, and so on. What, then, is the point of talking as though we have precise information about all this? Why not wait until we know a lot more about all this?

This is a strong and sensible objection, but I think it fails. There are several things I want to say in response.

First, I use conservative estimates to give critics of life extension the benefit of the doubt. I'll assume that the bad social consequences of widespread life extension (primarily Malthusian consequences) significantly outweigh the good consequences, even though I'm not convinced that this is true. I'll assume that the burdens of life as a Have-not reduce the value of a year of life by 50 percent even though I think there's no reason to believe that the burden of being a Have-not in such a world is anywhere *near* that bad. These are all very conservative estimates from the standpoint of someone arguing in favor of life extension, for they lend support to critics of life extension. If the harm to Have-nots is smaller, then the case for life extension is stronger. (It's hard to know which estimate of the percentage of Haves is the conservative one, for that percentage cuts both ways: If there are more Haves, there will be more welfare in that group, but there will also be more harm to the other two groups, and if there are fewer Haves, their total welfare—and the total harm they cause—is correspondingly less. I'll come back to this later.)

I will argue that even on conservative assumptions, it's likely that Promotion produces a lot more welfare than Inhibition. Thus, even if our figures are wrong, they are wrong in a direction that favors our critics. They have no reason to complain about our lack of precision, for they benefit from it.

Second, we don't have the luxury of avoiding a choice between the life extension research funding policies of Inhibition and Promotion. In a world where life extension research will be aggressively funded only if we decide to do so, not having a policy is tantamount to having a policy of *not* funding this research aggressively. It's a policy of Inhibition by default. It's true that we can't be sure the facts support Promotion, but we also can't be sure they support Inhibition, and yet we're choosing Inhibition right now. Because we can't avoid or postpone making a choice, we have to make a decision despite not having much hard information.

Third (and this point is related to the second), there's a distinction between (a) the strength of our reasons for being committed to a particular policy and (b) the strength of our evidence for the factual claims that support that policy. We can't be confident of the estimates we'll make about welfare, harm, the percentage of Haves, and so on, but when we have to decide and we have only very shaky information, our commitment to that decision need not be equally shaky. We can be confident that we've done our best to get the facts right (even if we can do very little right now) and that we've done the best we can to reach the right decision without being as confident about the factual claims that decision is based on.

Fourth, the numbers I'll toss around in the next section are not meant to indicate precisely what will happen. They are meant to state our intuitions precisely. They are meant to show what the world would *have* to be like for the advent of life extension to produce more welfare than a world without life extension, and the reverse. To put this another way, these numbers are expository devices. Here's a familiar example of using precise figures as an expository device. Scientists now believe there are roughly 100 billion stars in our own galaxy and that at least 17 percent of those stars have a planet roughly the size of the earth. Some scientists have used such figures to say something along these lines: if 1 in 1,000 of those earths has life, and 1 in 1,000 of those planets has complex life, and 1 in 1,000 of those has intelligent complex life, then there are at least 170 other intelligent forms of life in our own galaxy alone (and there are 100 billion galaxies in the entire universe . . .). The point of such calculations is not to estimate how many other intelligent species our galaxy contains, even though it may sound that way. The point is to dramatize just how vast the universe really is and how low the odds of forming intelligent life would have to be for us to be the only intelligent species in all the cosmos. Such calculations are a way of alerting us to the fact that even if the odds of intelligent life arising somewhere are vanishingly small, the universe is so vast that we can't rule out the possibility that there's someone else out there. The numbers merely help us see that more clearly.

The same thing is true of the calculations I'll offer in the next section. They are meant to show just what it would take—how much harm the Have-nots would have to suffer and how little gain the Haves would have to receive—for the advent of life extension to reduce overall welfare in the long run. As we will see from multiple examples involving somewhat speculative numbers, a world with life extension has less net welfare than a world without life extension only if the gain to each Have is rather small and the harm to each Have-not is stunningly large. We will also see that it's not very plausible that Have-nots will suffer that much harm per person.

I could, of course, make that same argument intuitively, without any numbers. The argument would go something like this: Suppose 10 percent of the population has a 1,000-year life expectancy. If someone's life span is a dozen times longer than normal, then it's likely to contain a dozen times as much welfare, or at least lot more than normal. Therefore, if even a small percentage of the human race gets such a huge gain in welfare, the total gain for all of them will be huge. The net harm all of this causes for each person who doesn't get life extension would have to be extremely large before it would be greater than all that welfare, and it's just not plausible that the harms will be that great. After all, how harmful would it be to live in a neighborhood where every tenth neighbor lives that long? Having a 768-year-old living next door isn't really a problem for you, and we should be able to handle the Malthusian challenge in the long run through Forced Choice or something similar. Moreover, in the long run, life extension can probably be made available to everyone, but this can't happen until we first develop the early, more expensive versions of it, so putting up with a span of time during which most people don't get life extension is necessary to get to a future where everyone gets it.

Now I want to state the forgoing argument more precisely. To do that, I'm going to make the same points using examples with estimated percentages of Haves and numbers representing total welfare or harm. Doing that will help us see more clearly why it's not plausible, in the long run, that the total harm suffered by Have-nots and Will-nots will be larger than the total gain in welfare for the Haves.

9.5 Maximizing welfare in the long run

Imagine two worlds with (just to keep things simple) 100 people each. I'll call the first world *the nonextended world*, for no one there has life extension, and the second world *the extended world*, for some people there have life extension (though others don't). For each world, we'll ask which population contains the larger total amount of welfare over some span of time. We need to make some assumptions.

First, we must estimate the life expectancy of the Haves and Have-nots. I will arbitrarily assume that the average life expectancy of each Have-not is 85 years. That's somewhat above the average life expectancy for developed countries right now, but let's assume that life expectancies continue to rise during the time that life extension is developed. As for the life expectancy of Haves, it depends on how effective a version of life extension we develop. Early versions may provide only an extra decade or two. Later versions might provide 30 or 40 extra years. However, if it's possible to slow aging, then in principle, it should be possible to completely halt it. I'll assume that we learn how to halt aging. Remember that if aging is halted, our life expectancy will be 1,000 years, according to the demographic estimate I discussed in section 1.7.

Next we must estimate the percentage of Haves. Over time, that percentage is a moving target. As incomes go up around the world and the cost of life extension technology falls over time (more of this later), the percentage of Haves will increase. Early on the percentage will be very small, and eventually, it will be much higher, but we have to pick a point in time to begin our analysis, so just to get started, let's pick the point where the percentage of Haves in our imaginary extended world is 10 percent.

Now we must estimate the harm suffered by the Have-nots and Will-nots in the extended world. To keep things simple, I'll assume that all Will-nots suffer as much as the Have-nots and simply speak of two groups: Haves and Have-nots. (In other words, the percentage of people who have access to life extension is larger than 10 percent, but some of those people turn it down so that only 10 percent of the population actually uses life extension.) This is another conservative assumption, for an average

Will-not really won't suffer as much as an average Have-not, all else being equal; this increases the estimate of total harm. Again, critics of life extension benefit from this assumption.

I will assume that the Have-nots suffer from Malthusian issues, the death burden, distress, and other factors and that they are harmed so badly that their welfare is *cut in half*. The Malthusian issues are tricky. On the one hand, I don't want to assume that we will impose Forced Choice and thereby avoid them indefinitely. There is a chance that we will not, and if we're estimating the total harm that life extension may bring, we can't ignore that chance. On the other hand, it's not safe to assume that we'll never institute something like Forced Choice. I suspect the most likely scenario is that population pressures may increase for a time before Forced Choice is instituted and effectively enforced over most of the world, so I will assume that in the near term—the time when the percentage of Haves is 10 percent, let's say—there are some Malthusian consequences, but they are not yet bad enough to make life unbearable. I'll also assume that by the time we impose Forced Choice, the world's population may be large enough that we suffer some crowding and other Malthusian issues even if we manage to prevent the world's population from getting any larger from that point on. In short, I'll split the difference and assume that there are some Malthusian harms, but they're kept to a manageable level in the long run. (If we fail to do that, then in the long run, life extension *will* do more harm than good.)

How can we estimate the amount of all those harms—the death burden, distress, Malthusian harms, and all the others? We don't have the kind of data here that we had about income and healthcare costs, so we'll have to lean a little harder on our intuitions about what's plausible. Here's a way to estimate those harms: compare a year of life as a Have-not in an extended world, where the Have-nots suffer various harms, to a year of life in a world where no one has life extension and those harms don't occur. The second year is better than the first. Obviously you would prefer the better year over the worse one. Perhaps you would even prefer the better year if it were less than a full year and the worse year was a full 52 weeks. Let's imagine that you have a choice: you can give up part of a better year in order to be free of the harms of the worse year. For example, you would certainly shorten your year by one hour to avoid the harms of 52 weeks as a Have-not in an extended world where your total welfare is cut in half. However, you probably wouldn't give up 11 months in order to avoid those harms.

At some point in the middle, however, you would be indifferent. In other words, there is some amount of time where you're indifferent between a short but better year of X length and a full but worse year. They are equally attractive or unattractive, and you don't care which you get—they have equal value to you. Suppose, for example, that you decide that 46.8 weeks (52 weeks minus 5.2 weeks) free of those harms is equivalent to 12 months of life as a Have-not in an extended world. You're indifferent between those two possibilities. If so, then you're discounting each year of life as a Have-not in an extended world by one-tenth. In effect, you think that the harm you suffer in a year of life as a Have-not in an extended world is equivalent to 90 percent of a year of life free of those harms. (Bioethicists will recognize that this is similar to a quality-adjusted life-year.[2])

How much of a year would you give up? It's hard to say. None of us have lived a year as a Have-not in an extended world, and individual decisions will vary. Speaking for myself, the closest I can come to getting a grip on this is to think about harms I know more about. Like most people, I've suffered depression, frustration, unhappiness, and psychological distress of various kinds at various times in my life. Perhaps those feelings are a lot like the distress many Have-nots will suffer in an extended world. As for the death burden (the extent to which your death is worse because you missed out on more time even though you live a normal life span—see section 7.3), it's hard to think of anything comparable in the world we know now. The closest I can get is to think about lost opportunities. Due to various life choices, I'm not precisely and exactly where I'd like to be. I missed some opportunities and made some bad choices. Like the death burden, these are purely comparative harms. I'm not worse off than I was, but I am worse off compared to how I might have been had I played my cards better or had better luck. As for Malthusian and other bad social consequences, I've spent time in crowded third-world cities that had vast slums, nondemocratic governments, and tiny economic elites. I can only imagine what it would be like to live there all my life and have no alternative, but I can at least imagine it.

Now let's add all that together. How much of a year of life could I lose and be indifferent between that reduced year free of unpleasant psychological states, regrets over significant lost opportunities, and living in a polluted, crowded urban setting and a full year with those harms? All I can do is guess, but for me, probably not more than 10 percent of a year. If those harms are comparable to what I would suffer as a Have-not in an extended world, then I have some reason to believe that my welfare as a Have-not in that world would be reduced by 10 percent.

But let's give opponents of life extension the benefit of the doubt, and work with a far, far more conservative estimate: 50 percent of a year. Let's assume that I would be indifferent between half a year free of all those harms and a full year that contained those harms. This estimate for harm to Have-nots is implausibly high; it's not likely that their lifetime welfare is cut in *half*. However, once again, opponents of life extension have no reason to complain about this assumption; it works in their favor.

So our comparison of an extended world with 100 people and a nonextended world with 100 people will assume that each Have has an average life expectancy of 1,000 years, each Have-not has an average life expectancy of 85 years, the percentage of Haves in the extended world is 10 percent, and each Have-not's welfare is reduced by 50 percent in the extended world. As I said at the end of section 9.3, we will use the number of years lived as a rough index of the total welfare for each person over the course of a life. We will express the welfare of a given world as the equivalent of a given number of life-years for the entire population of that world.

With all that in mind, let's compare the extended and the nonextended world. The nonextended world contains 100 people living 85 years each, so it contains the equivalent of 8,500 life-years of welfare (we'll leave their descendants out of this for the moment). As for the extended world, here is the total for the two groups of people:

10 Haves × 1,000 years = 10,000 life-years
90 Have-nots × 85 years × 50% = 3,825 life-years

This particular extended world has more welfare than the nonextended world by a margin equivalent to 5,325 life-years. In fact, even if we assume that the Have-nots are harmed so badly that they have no welfare at all, the world with life extension still has 1,500 more life-years' worth of welfare than the world without life extension (10,000 life-years in the extended world and 8,500 life-years in the nonextended world).

Just for further comparison, suppose that the percentage of Haves is only 5 percent, yet the Have-nots *still* lose half their welfare. In that case, the world with life extension has 9,037.5 life-years of welfare:

5 Haves × 1,000 years = 5,000 life-years
95 Have-nots × 85 years × 50% = 4,037.5 life-years

Again, the total welfare of the extended world is greater, this time by the equivalent of 537.5 life-years. Bear in mind that it's even less plausible that Have-nots lose half their welfare in that world than it was that they'd lose half in the world with twice as many Haves. A plausible harm figure would be less and the resulting welfare more.

We could dramatically reduce the percentage of Haves—drop it to 1 percent, let's say—but then the harm to Have-nots drops dramatically too. There would be no Malthusian crisis, and the death burden would be quite minimal. The Have-nots might still suffer significant distress and perhaps suffer from the fact that some entrenched members of the elite are living far too long, but overall, it is hard to say that their welfare is cut in half in such a world. Besides, it's not likely that only 1 percent of the population will ever be able to afford life extension; over time, that percentage is likely to increase.

We have been speaking of *total welfare*. This is the measure endorsed by total utilitarianism. The extended world also has greater welfare according to average utilitarianism. Here the analysis is much simpler. An extended life, especially one that lasts for centuries, has more total utility than a normal life span, even if the quality of life is lower than it would be in a normal life span (unless, of course, the quality is absolutely abysmal, but by then, one might well elect to go off one's life-extending meds and resume aging). Even if only a small percentage of the human race extend their lives, they will pull up the average for the entire population. Therefore, the extended world has a higher average welfare than the nonextended world.

By the way, when we think about a world with life extension, we have to distinguish between two ways of understanding average welfare that are not usually distinguished. In the preceding paragraph, I conceived of *average welfare* as average total welfare per lifetime: take the total welfare in each lifetime and average that across the population. However, there is another way to understand average welfare: average welfare per person per year. (Peter Singer noticed this distinction and articulates it as a distinction between quality of life and whether people have better lives.[3] We'll return to this in section 9.7.) If my arguments in chapters 2 and 3 are unsuccessful and critics of life extension are right to say that living as a Have will become less and less attractive with each passing year (thanks to boredom and other factors), then Haves may enjoy greater average welfare per lifetime but lower average welfare per year of life. However, that's another way of saying that my initial estimate of their total welfare is mistaken and that because their lives are not as good (on a year by year basis) as life in a non-extended world, I should not attribute 1,000 life-years of welfare to each Have. If you found my arguments in chapters 2 and 3 unconvincing and you believe that extended life would be a drag, then you can make suitable adjustments to my estimates and see what results. However, it takes quite a discount to make a difference. For example, even if the welfare of a Have's 1,000-year life is worth only 500 life-years (their welfare is cut in half too), then the extended world still contains more welfare than the nonextended world, albeit by a small margin: 8,825 life-years in the extended world and 8,500 life-years in the nonextended world.

You may object that this estimate is for a single generation of 100 people, 90 of whom live an average of 85 years each and 10 of whom live an average of 1,000 years. The second group far outlives the first group, so shouldn't we also consider every generation during those 1,000 years? If we do, don't all the generations of Have-nots have more welfare at stake than that first generation of Haves?

The answer is yes, but only if we compare just one generation of Haves with multiple generations of Have-nots. That comparison would be a mistake, for we should also assume that those 10 Haves produce children too and that each generation they produce gets life extension and has more children in turn. (If they don't, there's no risk of a Malthusian crisis, in which case the Have-nots would suffer much less.) If, for example, each pair of Haves has the same number of children as each pair of Have-nots, then there will be as many generations of Haves as there will be of Have-nots (though there will be more Haves hanging around, for they don't die off in 85 years). We could then count all the generations of Have-nots and all the generations of Haves who are born during a 1,000-year period and calculate the net lifetime welfare for all of them. However, the result of that calculation is merely a multiple of what we have already done. It's simpler just to compare a single generation of Haves with a single generation of Have-nots, and that's what we have done.

As I said, I've been working with conservative assumptions. Now let's remove the least plausible of those assumptions: the assumption that Haves are only 10 percent of the population. That will be true at a certain point in time, but not forever. The percentage of Haves will increase over time, partly because the cost of new technologies falls over time and partly because average living standards are rising around the world. These two trends will increase the percentage of Haves over time, and if they continue long enough, everyone in the extended world will have access to life extension.

How likely is this? Most new technologies start out expensive and later come down in price, often quite dramatically, especially if the market is very large and we find ways to automate the production. Consider pocket calculators, computers, or generic drugs whose cost of development was paid for some time ago—they all dropped dramatically in price over time. For example, the price of a television set immediately after the Second World War started at $499 for a 10-inch screen and went as high as $2,495 for a 20-inch model. Adjusted for inflation, the cheaper set would now cost $4,825.37 and the expensive one $24,126.85—and that's not even in color. Prices soon plummeted, and by the late 1960s, a single person on modest wages in a developed country could afford a color TV with an even larger screen.[4] Now many third-world households have them.

As for standards of living, they have been rising around the world for some generations now (albeit more in some parts of the world than others and with some backsliding here and there).[5] GDP per capita increased almost four times over from 1950 to 2010 alone, and during that period, global life expectancy at birth increased by roughly 20 years.[6] Economic historian Angus Maddison, associated with the University of Groningen and founder of the Groningen Growth and Development Centre, led a team of economists in estimating GDP per capita over time, adjusted for inflation.[7] According to their analysis, GDP per capita per year rose over time as follows:

Year 1	$467
1820	$666
1900	$1,261
1958	$2,607
1988	$5,056
1998	$5,709
2008	$7,614

According to Maddison, per capita GDP worldwide tripled in the half century from 1958 to 2008 and increased sixfold in a little more than a century, from 1900 to 2008.

Here is more evidence of rising living standards: according to the World Bank, the percentage of the world's population living in "extreme poverty" has dropped from 35 percent in 1992 to 14 percent in 2011 (the most recent year for which the World Bank has an estimate). More than 12 million children died before the age of five in 1990; the number is now half that figure.[8] The World Bank also reports that in 2012, 12.73 percent of the world lived on $1.90 per day or less; this is down from 44.1 percent (in inflation-adjusted dollars) in 1981.[9]

If these economic trends continue and the cost of life extension falls over time, then eventually it will be possible to make life extension available to everyone on earth. Therefore, when we estimate the total (and average) welfare of a world with life extension, we must consider scenarios where everyone who wants access to life extension has it.

Would everyone want life extension? For our purposes, it doesn't really matter. If we suppose that most people will not want it, then we must also assume that the Malthusian problems will not be as severe (or may not happen at all), that those who don't want extended life will also feel no distress about it, and that if they suffer a greater death burden, they won't mind that either. Moreover, those who do extend their lives will still add significantly to the welfare of that world, so the extended world will still have greater total utility than the nonextended world. However, it's safer to assume that most people will want extended life, especially once it becomes commonplace and familiar. That world will have more welfare than a nonextended world on both the total utilitarian and the average utilitarian approaches.

Suppose, for example, that life extension is available to everyone, that 90 percent of our population of 100 people extend their lives, and that the Will-nots find their welfare reduced by half (another highly conservative assumption). On those assumptions, we get this scenario:

90 Haves × 1,000 years = 90,000 life-years
10 Will-nots × 85 years × 50% = 425 life-years

The total welfare here is more than 10 times larger than the 8,500 life-years' worth we attribute to the nonextended world. Even if we assume that the quality of extended life is reduced by half as well, we get a total of 45,425 life-years, which still dwarfs 8,500 life-years.

Let me touch on a technical issue in utilitarian theory before we move on. When I talk about welfare (utility), I take an impersonal view. According to this view, a state of affairs with more welfare is better than a state of affairs with less welfare, even if no one who exists in the first state of affairs also exists in the second state of affairs, and vice versa. On this view, it is, for example, better to bring a healthy child into existence than a sickly child, even if we are talking about two different possible children. Creating the healthy child does not make the potential sickly child any better off, but impersonal theories of welfare say we should create the healthy child instead of the other one anyway.

However, there is also a person-affecting view of welfare, which says that one state of affairs is better than another only if it is better for someone—in other words, if there is at least one person who exists in both states of affairs. On this view, there is nothing wrong with creating the sickly child. That result conflicts with some common and strongly held moral judgments, and many philosophers reject the person-affecting view for just such reasons. But let's entertain the person-affecting view for a moment.

If we create a world where life extension is available, then, for reasons familiar to students of the nonidentity problem, we will affect who comes into existence over time. If, for example, there had not been a peacetime draft in the early 1950s, my father would never have met my mother at the USO club where she worked and neither I nor this book would be here. Even as minor a policy change as altering the speed limit will affect reproductive behavior (you get home to your spouse half an hour later, and the baby you make that night comes from the same egg but a different sperm—an entirely different baby than the one you would have made had you come home a bit earlier). If raising the speed limit can do that, imagine what a major change like the advent of life extension might do.

Now compare an extended world to a nonextended world. At the point when life extension first becomes available, both worlds contain the same people. (They are two possible futures for the same world.) Over time, most people who exist in one of those worlds will not exist in the other, and vice versa, and eventually the populations will be entirely different. They will contain some of the same people only in the first generation or two, or perhaps the first handful of generations. Once the two populations have completely diverged, life extension does not make anyone better off than that person would be in the other possible world. According to the person-affecting view, that extended world is not better in any way.

Therefore, if the person-affecting view is correct, my argument that the extended world has more welfare works only if we suppose that the total gain in welfare for the first few generations of Haves exceeds the total harm to the first few generations of Have-nots. For reasons illustrated by the comparisons of extended and nonextended worlds earlier, I believe this will be so, but no one can be certain.

In any case, I share the reservations of other philosophers about the person-affecting restriction. My biggest reservation is that I have a strongly held moral judgment that a long-range policy that increases welfare over time is a good thing *because* it increases welfare, all else being equal, even if it also affects reproductive behavior and eventually produces a different population. The person-affecting view implies that that judgment is incorrect. I cannot give up that judgment, so I reject the person-affecting view. I therefore use an impersonal approach in this chapter.

9.6 How to argue that a world without life extension has greater net welfare than a world with it

What would it take to get the result that a world *without* life extension has more welfare than a world with life extension? (This is the counterpart to the question where we ask how low the percentage of earthlike planets with life would have to be—and how low the percentage of planets with complex life would have to be, and so forth—for the human race to be the *only* intelligent life in the cosmos.) To get the result that there's more welfare in a world without life extension, we must keep the harm per Have-not high while reducing the number of life-years the average Have will get so that Promotion produces less than 8,500 life-years. Here is one scenario where that happens:

10 Haves × 300 years = 3,000 life-years
90 Have-nots × 85 years × 0.5 = 3,825 life-years

That worked: now Promotion produces 6,825 life-years, 1,675 fewer than Inhibition.

But consider what it took to produce that result. We had to assume that letting one person in ten live 215 years longer than other people will reduce the quality of life for the other nine by *half*. We could tinker with these figures a bit, but if we reduce the burden on Have-nots to something more plausible, the net welfare in a world with life extension goes up. Any estimate that tells us there is more welfare if life extension is never developed must assume a huge amount of harm for each Have-not and only modest gains for Haves. It's very hard to argue that this is likely, for most of the harms to Have-nots turn on the number of Haves and how long they live. The less the benefit per Have or the smaller the number of Haves, the less the harm to Have-nots.

Moreover, it's not plausible to assume that the Haves will never live more than 300 years, not just in the coming decades, but in the next century, or the century beyond that, or 1,000 years from now, and so on. In other words, we have to assume that it is simply impossible to extend human life much more than that, no matter how advanced our medical science and technology become. (This is one reason I assumed that Haves would have a 1,000-year life expectancy.) Given that DNA was discovered less than a lifetime ago and we have now mapped the entire human genome and begun to develop an effective technique for genetic engineering, it's just not realistic to think that life extension technology will be frozen forever at the level of modest gains. It is more likely that sometime in the future, we will learn how to completely halt, prevent, or reverse aging and that life extension will bestow a 1,000-year life expectancy. Again, we want to know all the consequences of making life extension available and that includes consequences a long, long time from now.

Finally, it's not plausible to assume that it will always be fiscally impossible to provide life extension to most of the human race, not just in the next 100 years but forever and ever. In the long run, standards of living are rising around the world, and it's likely that that will continue. It may even happen that artificial intelligence, automation, and robotics will lead to an economy largely divorced from human effort so that we're all supported by a vast technological system capable of generating enormous prosperity while most of us enjoy a life of leisure. (This will cause problems of its own, including a wrenching transition to that kind of society and the problem of finding meaningful lives without work, but that's a topic for another book.)

I conclude that in the long term, a world with life extension is highly likely to have greater total (and average) welfare than a world without it.

9.7 Peter Singer's objection

In the previous section, I assumed that Forced Choice would be imposed soon enough to avoid a Malthusian disaster but not soon enough to prevent *all* damage caused by population increase. Of course, that may be too pessimistic; perhaps the Malthusian issues will not be serious, or perhaps we'll impose Forced Choice sooner than I anticipate. However, Peter Singer has argued that a world with life extension will have either the same or less total welfare than a world without life extension *even if* we successfully limit our population so well that the extended world has no more people than the nonextended world.

Singer believes (as I do) that if life extension becomes available, it will be necessary to impose controls on reproduction to prevent a Malthusian crisis. He compares two possible worlds. In one of them, everyone has life extension, but something like Forced Choice is imposed. In the other, life extension is not available. Singer then makes a critical assumption: both the extended world and the nonextended world have the same number of people, presumably because the controls on reproduction are meant to ensure that the world with life extension doesn't have a larger population than the world without life extension.[10]

Consider a very simple version of this comparison. Let's suppose that in the nonextended world (the one where no one has life extension), people have an average life expectancy of 85 years. Let's also suppose, just for discussion, that in the extended world, people live twice as long (this was Singer's assumption)—170 years. Because the extended world imposes controls on reproduction, both worlds will have the same number of people at any given time. This means that both worlds have the same total number of life-years, but the nonextended world will have twice as many people during any 170-year period:

Extended world: 100 people living an average of 170 years each; 1,700 life-years

Nonextended world: 200 people living an average of 85 years, so 100 people at any given time; 1,700 life-years

In short, the total number of life-years (our index for quantifying welfare) will be the same in both worlds even though each Have gets twice as many years as each Have-not. Therefore, the extended world and the nonextended world have the same total welfare not only at a given time but forever. Therefore, a world with life extension will not have more welfare than a world without it. (The same is true of worlds where Haves live 1,000 years, or any other life span, compared to nonextended worlds.)

In fact, Singer tells us, there is good reason to think the extended world may have *less* total welfare. People who live extended lives may well find that each additional year contains less welfare than it did earlier in life, when everything was new. Why? Presumably Singer is sympathetic to the criticisms of extended life that we examined in chapters 2 and 3 and believes we would become bored or life would have less meaning, and so forth. Therefore, there's less total welfare in the extended world, for that world has the same number of life-years but the welfare per year is less (it declines over time).[11]

There are two ways to evaluate Singer's argument. First, we can evaluate his philosophical arguments. For example, we might start with the fact that Singer appears to subscribe to total utilitarianism and argue that total utilitarianism is a mistaken view or even argue against utilitarianism altogether. Second, we might focus on the factual claims Singer assumes and question whether those claims are plausible. Rather than open a long and well-worn philosophical controversy over the merits of utilitarianism in general and total utilitarianism in particular, I'll give Singer the benefit of the doubt on his philosophical claims and evaluate his arguments in the second way, focusing on his factual claims.[12] Singer's objection to life extension depends on two factual claims: (a) average welfare in the extended world is lower than average welfare in the nonextended world and (b) both worlds have the same population at any given time.

Both of these factual claims are questionable. We'll start with the first. We can't safely assume that the average welfare per life-year in an extended life will be lower than average welfare per life-year in a normal life span, nor does Singer really argue for that claim. In chapters 2 and 3, I tried to suggest some ways in which we might avoid boredom and find new sources of satisfaction over the course of an extended life. If I'm right, then the quality of life in an extended life could be the same as that in a normal life span. It could even be greater if we use our extra time to learn how to live better. Given enough time, we might enlarge our spiritual and psychological horizons. Moreover, if we halt aging entirely, then people will be spared the indignities and discomforts of growing old, and death will be a more remote worry. Singer's first factual claim is not established and may well be false.

But the real problem is with Singer's second factual claim. It's very likely that a world where life extension is widely available will have a larger population than a world that has no life extension, even if reproduction in the extended world is controlled by Forced Choice or something similar. There are several reasons for this. For one thing, it will take time to develop and impose a policy of Forced Choice. People don't take kindly to being told how many children they may have, so such a policy will not be popular, and there will be little public pressure to institute and enforce such a policy until some Malthusian trends begin to appear (just as worldwide political will for measures to fight climate change did not appear until dangerous climate trends had gone far enough to become obvious to many people). Instituting Forced Choice will likely take at least two or three generations, and during that time, population will already be increasing beyond what would happen if life extension were never developed in the first place, even if the controls on reproduction work so well that the population size stabilizes from that point on. By the time Forced Choice is imposed, some of the damage is already done, so to speak.

Let me now respond to Singer's factual claim about population sizes from another angle. Comparing a world with life extension to a world without life extension, as Singer does, is the relevant comparison if we want to know whether a world with life extension is a better world. Of course, we do want to know that. However, we also want the answer to another question: Which life extension research funding policy should we adopt? Our choices are Promotion, which funds life extension aggressively enough to enable it to proceed as fast as science can make it go, and Inhibition, which either does not fund it at all or funds it at some lower level so that the advent of life extension is delayed. If we're deciding which funding policy to adopt, then the relevant comparison is between a world where life extension is developed sooner and one where it's developed later. The reason for this is that it's probably impossible to prevent life extension from being developed sooner or later, especially since geroscience is already so far along. Prohibition—the policy of preventing life extension from being developed and used at all, forever—is simply not a practical alternative, regardless of its interest for moral theory.

Therefore, we should compare these two worlds: Promotion World, where life extension is developed faster and made widely available faster, and Inhibition World, where life extension is developed and made available much later. If life extension is inevitable in the long run, as I believe, then these are the two relevant alternatives.

In Promotion World, life extension becomes available sooner, so the time when life extension pushes population trends upward (or at least reduces their decline) will come sooner as well. In Inhibition World, these developments will come later. Suppose, for example, that in Promotion World, the point where life extension begins to push population trends upward (or at least reduce their decline) comes in the year 2100, while in Inhibition World, that point is not reached until later, in 2200 (let's suppose). Now suppose that Forced Choice is imposed in both worlds, with similar results, and that in both worlds, the population eventually levels off at some tolerable level. Suppose further that it levels off at the same population level in both worlds: 10 billion people, for example. Which world will have a larger population from now until the end of the human race, whenever that comes? You might think the answer is "neither—they have the same population." However, all else being equal, the world with the larger population is the world where population trends inched upward sooner. Promotion World sees that increase a hundred years sooner than Inhibition World, so the total number of life-years over time is greater in Promotion World. During that first 100 years, Promotion World has marginally more people than Inhibition World, so even if the populations thereafter are the same size, Promotion World has marginally more total welfare over time. Therefore, Promotion would probably produce more welfare than Inhibition would.

Singer might be right about scenarios where the extended world and the nonextended world have the same size population, but that scenario is highly unlikely to occur, and his conclusions don't apply to scenarios where the population of an extended world is larger than the population of a nonextended world.

9.8 Conclusions

Y. The best way to decide which life extension funding policy is morally most justified is to apply a set of midlevel moral principles that call for maximizing welfare whenever doing so either does not violate a right or violates a right but the amount of welfare is very, very large compared to the importance of whatever interest the right protects without violating any rights. (Section 9.2)

Z. In the long run—and quite possibly in the short run too—a world with life extension will almost certainly have greater welfare than a world without it. (Sections 9.3 through 9.6)

AA. It's very likely that a world with life extension will have a larger population than a world without it and have greater welfare than the world without it. (Section 9.7)

10   Deciding among the Groups—Which Rights Are Relevant?

10.1 Introduction

In chapter 9, I argued that Promotion produces the most welfare, that a world where life extension is available will probably have greater net welfare than a world where it is not. However, there's more to moral theory than welfare. There are also rights, and sometimes maximizing welfare involves violating rights. In this chapter, I will focus on three rights: a right to equality, a right to self-determination, and a right against harm. The members of each of the three groups—Haves, Will-nots, and Have-nots—have all three of these rights. However, as we'll see, not all of those rights are relevant here.

In this chapter, I will argue for three conclusions. First, the right to equality does not weigh in favor of either Inhibition or Promotion. The right to equality must be respected, but the importance of equality is not a reason against Promotion even if we have good reason to believe the right to equality will be violated once life extension becomes available. Second, the right to self-determination does not weigh in favor of either Inhibition or Promotion, for neither policy violates the right to self-determination of any of our three groups: Haves, Have-nots, or Will-nots. Thus, the right to equality and the right to self-determination are not relevant here. Third, the right against harm *is* relevant, and it favors Inhibition. Promotion will harm the Have-nots and Will-nots, but—I will argue—Inhibition will not harm the Haves. Inhibition prevents the Haves from becoming better off, but that's not a harm.

10.2 Rights and welfare

We'll begin with a brief overview of rights and their relation to welfare.

Sometimes maximizing welfare seems unjust. For example, we might maximize welfare by killing one healthy innocent person and taking his organs to save three other people who will die without them (imagine they're too low on the waiting list to get a transplant any other way), thereby killing one person to save three others. Almost no one believes that's morally right. To take another example, we might maximize welfare by passing a law that says all people ages 20–22 must perform two years of "national service" for free (room and board provided by the government), during which they must work on public construction projects, thereby saving everyone over 22 a lot of money. Such a law strikes me as unjust.

So there is more to morality than simply maximizing welfare. Many philosophers express this idea by claiming that there are moral rights. When we say that people have a right to something, we're saying (whether we realize it or not) that they have an interest or choice of such importance that we should give that interest or choice special priority and protection. Moreover, we give the things the right protects greater moral weight than the moral value those things would have if we considered them merely as good or bad consequences apart from their relation to some right. For example, most of us believe that people have a moral right to free speech, even if the speech is false, disturbing, or harmful (within limits and with exceptions, such as shouting "Fire!" in a theater, passing military secrets to the enemy, slandering others, and so on). We believe, at least implicitly, that the value of free speech, both to the speaker and to others, is usually morally more important than the harm such speech might cause.[1] In addition, most (perhaps all) rights have corresponding duties. If, for example, you have a right not to be harmed, others have a duty not to harm you. As a result, sometimes we have a duty to do something that conflicts with maximizing welfare.

So morality requires that we maximize welfare so long as we don't violate rights by doing so. However, we don't want to say that we can *never* maximize welfare at the expense of violating a right. Consider an example involving the right to self-determination—roughly speaking, a right to make decisions about our own lives. Suppose that one person in the entire human race has a genetic mutation that produces a protein that makes her immune system twice as effective as normal. Leaving aside what's legally possible, wouldn't morality require (or at least permit) taking a small sample of her blood so we can clone the blood cells and produce a serum that saves hundreds of millions of lives, even if she adamantly refuses to consent to this? You may be an absolutist about rights, but I suspect most of us would say that maximizing welfare takes priority over respecting her right in that case. In my judgment, morality requires taking a blood sample from her by force, if necessary. Cases where morality requires maximizing welfare even at the expense of violating rights should be rare, but it's not plausible to say there are none at all. That's why our theory allows us to maximize welfare by violating a right when the amount of welfare at stake is so large that it outweighs the right. (Later on, we'll talk about how to determine *when* welfare outweighs a right.)

To sum up, our moral theory requires us to maximize welfare except when doing so violates a right, unless the amount of welfare at stake is so large that it outweighs the moral importance of the interest protected by that right. For the rest of this chapter, we'll ask whose rights, and which ones, are violated when we maximize welfare by making life extension available.

10.3 The right to equality favors neither Inhibition nor Promotion

I argued in chapter 8 that people have a right to equality. I also argued that unequal access to life extension is unjust and that the Have-nots have a right to subsidized life extension. Of course, the Haves and the Will-nots have a right to equality too; everyone does. However, in this section, I'll argue that neither making life extension available nor failing to do so violates anyone's right to equality. To put this another way, the right to equality does not count for or against making life extension available. To see why, let's briefly review some arguments from chapter 8. After describing a world where some get life extension and some do not, I argued that the big question is not whether such a world is morally unattractive. The big question is what to do about it.

I made three egalitarian arguments that unequal access to life extension is unjust. First, it violates the right of equal opportunity. Second, it magnifies other unjust inequalities. Third, most theories of equality (and some alternatives to egalitarianism that have egalitarian implications, like prioritarianism and sufficientarianism) imply that this particular inequality is unjust. Then I argued that Haves have a duty to subsidize life extension for the Have-nots because it's wrong to do nothing about an unjust inequality and it's also wrong to deal with inequality by practicing collective suttee (also known as leveling-down) by denying extended life to everyone. The only other thing Haves can do about it is to subsidize as many Have-nots as they can. Therefore, Haves have a duty to do just that. Because they have that duty, the Have-nots have a right to equality—in practice, a right to such a subsidy.

Then I addressed this question: What if we can be pretty sure that many or even most Haves will not subsidize life extension for the Have-nots, either because they're too stingy or because it's not fiscally possible for the Haves to provide life extension to all the Have-nots? Is that a reason for Inhibition? I gave four reasons why it's not. I'll briefly summarize them again here.

The first and primary reason why an expectation of inequality is not a reason for Inhibition concerns equality and leveling-down. As I argued in section 8.3, Inhibition is a kind of leveling-down, and equality doesn't require leveling-down. There's a virtually universal consensus to the effect that equality doesn't require leveling-down, and egalitarian philosophers acknowledge this. Instead of arguing that leveling-down is morally acceptable, they argue that egalitarianism doesn't require leveling-down (or that some moral principle forbids it and overrides egalitarianism), thereby implicitly agreeing that leveling-down is wrong. Because the right to equality does not require leveling-down to correct breaches of that right, it also doesn't require leveling-down to *prevent* breaches of that right. Therefore, even a high probability that Have-not rights to equality will be violated is not a reason for Inhibition. It is a reason for doing other things about the inequality, but stifling the technology that we anticipate will be distributed unequally is not one of them.

Second, some Haves are generous and will subsidize life extension for the Have-nots; inhibiting life extension prevents stingy Haves from breaching their duty of equality but also prevents generous Haves from performing that duty, so it prevents breaches at the cost of preventing performances too. Inhibiting the development of life extension so that stingy Haves will not be in a position to breach their duty to subsidize life extension for the Have-nots is like outlawing medical care because stingy Haves refuse to provide it to Have-nots. There is something to be said for punishing those Haves, but preventing the means of performing the duty does not prevent breaches of the duty except in a narrow, technical sense.

Third, we cannot justify this as punishment for stingy Haves, for we can't punish someone for something they haven't done (it's in the nature of Inhibition that this punishment would preclude the crime). Moreover, some Haves are not stingy; we can assume that some will donate money to funds for Have-nots and that some governments will subsidize life extension for at least some Have-nots. Preventing life extension from being developed does more than punish stingy Haves; it also punishes generous Haves and the Have-nots those generous Haves would otherwise help.

Fourth, inhibiting the development of life extension delays the day when life extension may be cheap enough and the world's economy prosperous enough to provide life extension to everyone, thereby hurting still more Have-nots.

So the right to equality is not a reason for Inhibition. However, it's not a reason for Promotion either. The right to equality tells us that *if* the Haves get access, they must try to provide access to the Have-nots too. It does not tell us that we must develop life extension so that we can then distribute life extension equally any more than the right to equality requires that we develop faster speeds for internet access so that we will then be able to provide equal access to that technology. There are other good reasons for developing something like that, but doing this in order to create a new opportunity for equal distribution is not one of them. Therefore, the right to equality does not count for or against Promotion, and we can set it aside.

10.4 The right to self-determination favors neither Inhibition nor Promotion

It's commonly agreed that people have a right to self-determination—roughly speaking, a right to run your own life as you see fit.[2] The right to self-determination is not about what is good for you or even about what you think is good for you; it's a right to determine the nature and course of your own life however you choose, for whatever reasons strike you as good ones. Usually, your reasons concern your own welfare, and you are usually right about what promotes your welfare, but your right to determine your own life is not contingent on your being right about what promotes your welfare or even contingent on you *trying* to promote your welfare.

There are some qualifications and refinements. Your efforts at self-determination may not harm others. There's a boundary—hard to draw in practice—between what, to put it crudely, falls within your life and what does not. It is generally thought that you can exercise this right only if you are competent to do so and that others may interfere with your decisions if you are not competent or if your decision is not well-informed (such that your action won't achieve what you think it will).

The most important refinement for our purposes is that the right to self-determination is generally thought to be a *negative* right (a right against interference), not a *positive* right (a right to assistance). The right to self-determination requires us to respect the self-determining choices of others by refraining from interfering with them, but it typically does not require us to help them get what they want. If the right to self-determination were a positive right, we would all have the right to demand help from anyone else whenever we need it to determine our own lives as we wish, and they would have the right to demand the same from us. Such an expansive right to self-determination would ultimately undermine self-determination. This is not to say that we never have a positive duty to help others determine their lives, only that such a duty arises only in special cases, including but not limited to these: when we agreed to help one another or made some other commitment to do so or when we have a special responsibility for someone else (as parents do for their children).

Thus, the Haves, Have-nots, and Will-nots all have a right to determine the course of their own lives, subject to those qualifications. The Haves want to extend their own lives, the Will-nots (or some of them, anyway) want to live in a world without life extension, some Have-nots will want access to life extension, and other Have-nots will want life extension to go away. It seems like the right to self-determination pulls in two directions; to some extent, it's a reason for Promotion and, to some extent, it's a reason for Inhibition. However, it only seems that way. As it happens, neither Promotion nor Inhibition violates the rights of self-determination of any of these three groups.

The self-determination of Have-nots and Will-nots

Let's start with those who prefer to live in a world where life extension does not exist: the Will-nots and some Have-nots. Does the right to self-determination cover that preference? The right to self-determination is not a right to have what you want; it's a right to make decisions and take actions about matters that are, roughly speaking, your own business and no one else's. A preference for living in one kind of world rather than another is not an action or decision of that kind. It's true that some intentions can be fulfilled only under certain conditions, but it doesn't follow that a society that does not provide those conditions has thereby interfered with that person's right to self-determination. I prefer to live in a world where a large percentage of the population is interested in intellectual pursuits, but people who are indifferent to such things are not violating my right to self-determination.

You might think that living in an extended world does interfere with the self-determination of Have-nots and Will-nots. It means, for example, that they can't get the opportunities they want (the Haves are hogging all the senior positions) or afford real estate as easily (the world is more crowded and the Haves bought all the land long ago), and so on. My answer is that these things may happen, and they are serious grievances. However, these things are covered by the right against harm, not the right to self-determination. There are many situations where I can do things under one set of social conditions that I can't do under another set, and I prefer the first set. However, bringing about the second set of conditions does not thereby violate my right to self-determination. For example, I wish the government would fund social security more generously (by taxing the rich more heavily) so that I could pursue other projects that are possible only if I spend money that I now set aside for retirement. However, society's failure to do so is not an interference with my self-determination. It's true that in an extended world, Have-nots and Will-nots are forced to endure various things: distress, increased death burdens, Malthusian consequences, reduced death benefits, and so on. Once again, however, those interests are covered by the right against harm, not the right to self-determination.

Here is another way to argue that Promotion interferes with self-determination. Imagine a librarian who allows a study area to be noisy and disruptive, and it's been that way for years. We might well say that the librarian's failure to straighten things out is an interference with the self-determination of those who want to use the study area for study. Here there is a situation that is not as it should be and someone who is failing his or her duty to set things right. That counts as an interference with self-determination. Does permitting life extension to become available to Haves interfere with the self-determination of Have-nots and Will-nots in this way?

It does only if there's a reason to think that a world with available life extension is not legitimate and that a world without life extension *is* legitimate. (It also requires that there be someone who has a duty to set things right.) However, this argument requires some *other* reason to think that Promotion is wrong. We can't say that Promotion is wrong because it interferes with their self-determination, for that argument becomes circular; it amounts to arguing that Promotion interferes with their self-determination because it is wrong, and it is wrong because it interferes with their self-determination. Whatever other reason we might have for thinking that Promotion is wrong can stand on its own. We might argue that Promotion is wrong because it violates the Have-nots' and Will-nots' rights against harm, but we should not count that rights violation twice by describing it under two different names. Every harm interferes with self-determination in some sense, but we do not count all harms as two rights violations: a violation of the right against harm and a separate, additional violation of the right to self-determination.

The self-determination of Haves

What about the Haves? The Haves intend various things we can classify as efforts at self-determination, and some of these things require the extra time life extension would give them. They will pursue whatever projects they have now over a longer period of time or take up new projects. They may pursue new careers, new experiences, new relationships—or continue all the old ones over a longer span. Living life involves countless efforts at self-determination, and gaining additional years of life opens up time for additional efforts at self-determination.

However, the same reasons to think that Promotion does not violate Have-not or Will-not rights to self-determination also tell us that Inhibition does not violate the self-determination of Haves. (I mean potential Haves, obviously.) Again, society's failure to provide the conditions necessary for certain acts of self-determination is not necessarily a violation of the right to self-determination. A Have's right to self-determination does not imply a right to have life extension aggressively funded. And again, there is a circularity in arguing that Inhibition violates the Haves' right to self-determination because Inhibition is wrong if the reason it's wrong is that it violates the right to self-determination. And once again, if there is some other reason to think that Inhibition is wrong, that reason can stand on its own without being double-counted as a violation of the right to self-determination alongside whatever other rights it might violate.

There is another reason to think that Inhibition does not violate the Haves' right to self-determination. This reason is unique to the Haves; it doesn't parallel the arguments concerning the self-determination of Have-nots and Will-nots. Rather, it turns on the fact that, as I explained above, the right to self-determination is generally understood to be a negative right (a right against interference), not a positive right (a right to assistance). For example, I have a right to try to write another book and others have a duty not to disturb me when I'm writing, but no one has a duty to edit the manuscript or help me publish it. In particular, no one has a duty to fund a year free of teaching so I can work on it. Other people have a duty not to intercept your application for a job, but they don't have a duty to drive you to your job interview.

For the same reason, our current funding policy does not interfere with life extension research. It does not take the form of blocking research funding from being provided, or destroying someone's research records, or banning the research. No one's negative rights are being interfered with here. Instead, it consists of not funding such research as much as society could—a failure to provide help. The duty to avoid interfering with self-determination is not (absent some special duty or an illegitimate status quo) a duty to provide help, so failing to help does not violate that duty. Moreover, this is probably happening mostly through ignorance or indifference or simply because those who direct research funding are focused on other things they consider more important or more practical. Society is not interfering with the Haves; it's merely choosing not to fund life extension research as aggressively as Haves might like. Thus, the fact that lack of funding for life extension research prevents the Haves from determining their lives in some ways or to some extent is not thereby a violation of their right to self-determination even though their efforts at self-determination would be more successful if such research were funded more generously.

10.5 The right against harm favors Inhibition

Unlike the right to equality and the right to self-determination, the right against harm is relevant to the question of whether we should make life extension available. However, as we'll see, that right cuts in only one direction: making life extension available will harm the Have-nots and Will-nots, but failing to make it available will not harm the Haves. (We'll consider the Will-nots and Have-nots and their right not to be harmed in the next chapter.) The reason for this is that although failing to make life extension available will prevent the Haves from gaining vast amounts of welfare, preventing someone from gaining a benefit is not a harm per se. Being harmed means you are worse off; failing to be better off does not usually count as a harm. This claim is consistent with the three major theories of harm: comparative theories, noncomparative theories, and theories that combine comparative and noncomparative elements.

According to noncomparative accounts, you are harmed if something causes you to be in a state that is bad for you in a noncomparative way. There are many versions of this idea: the state may be intrinsically bad for you, or it may fall below some normatively set threshold of well-being, or it may be a state that you can rationally desire not to be in.[3] However, none of this seems true of having a biologically normal life span. Having a normal, nonextended life span is not intrinsically bad, nor does it push you below some threshold of well-being no one expects to fall beneath. You can rationally desire to have extended life (if my arguments in chapters 2 and 3 are sound), but if the fact that desiring some improvement in your condition is rational means you are harmed when you don't get that improvement, then we're all harmed in countless ways just because we can rationally desire to be better off. That's far too expansive a concept of harm. Therefore, according to noncomparative accounts of harm, Inhibition does not harm the Haves, for it leaves them with the same normal life span everyone always had.

According to comparative theories of harm, something harms you if it makes you worse off than *you* would otherwise have been.[4] According to such theories, you might, conceivably, be just as well off as you were before—or even better off—but still be harmed because something left you less well off than you would have been otherwise. For example, if someone destroys a will under which you would have inherited millions of dollars from your rich uncle, you've been harmed even if, for other reasons, you're wealthier now than you were before the will was made.

The Haves are not worse off in this sense unless there is reason to think that they are left worse off than they would otherwise have been. However, the better off condition they would "otherwise" have enjoyed cannot simply amount to the fact that there is some alternate possible world where they would be better off. If the possibility of being better off is all it takes, then we are all harmed whenever our lives are not as good as they might be if others behaved differently. The sense in which we would otherwise have been better off has to be something like an entitlement or a reasonable expectation, like the kind you have under your uncle's will. To argue that Inhibition harms the Haves in this way requires showing that they are entitled to Promotion. At this point in the discussion, it's still an open question whether justice requires Promotion, but even if it does, we cannot argue this way: The Haves are entitled to Promotion because Inhibition harms the Haves, and Inhibition harms the Haves because they would be better off under Promotion. That's blatantly circular.

The Haves are not worse off under Inhibition than they were before; they just don't get better off. Their life spans do not change for the worse; they just don't change for the better, and they still live as long as everyone else. Therefore, under comparative and noncomparative theories of harm, Inhibition does not harm them, and what is true of both theories is also true of theories that combine comparative and noncomparative approaches. Therefore, Inhibition does not violate their right against harm.

You might object that suffering pain and indignity from various age-related diseases and other maladies associated with old age is a harm. Arthritis, cancer, and other conditions that correlate with age make people worse off and surely count as harms. Life extension enables the Haves to avoid these things. Doesn't that mean that life extension enables the Haves to avoid harm and therefore Inhibition harms the Haves? (Here we're talking about the harms of the aging process itself, not the harm of dying sooner than one might.)

We can, of course, weigh all that into the balance. However, I suspect that the total amount of harms stemming from senescence that the Haves suffer under Inhibition is less than the total amount of harms suffered by all the Have-nots in an extended world: distress, Malthusian consequences, and all the rest of it. Moreover, because there are probably more Have-nots than Haves (for at least a generation or so), the total harm suffered by Have-nots in an extended world exceeds the total harm suffered by Haves under Inhibition. This is all pretty speculative, of course, but my critics should accept these concessions; they too are likely to claim the Have-nots suffer more harm from life extension than the Haves suffer from not having life extension. When you can't help harming one party or another, you violate a right to be harmed no matter what you do. It makes sense in that situation to accept the lesser amount of harm (to Haves) and thereby avoid some larger amount of harm (to the Have-nots and Will-nots). For this reason, the harms of senescence suffered by Haves are not enough to tip the scales in favor of Promotion.

And there is another reason not to say that Inhibition harms the Haves by depriving them of the means to avoid the harms of senescence. There are many situations where you might avoid some harm if others gave you, for example, more money, more opportunities, or more help of some other kind. However, failure to give you that help is not thereby a harm to you. You could help me avoid the pain of age-related arthritis by paying for an extremely expensive anti-inflammatory drug that will make that pain go away. However, you haven't *harmed* me by refusing to pay for it.

We could try to argue that the Haves have not only a right against harm but also a right to be helped. Some say that we have a positive right, a right to be helped, when the burden of helping us is very small compared to the harm we will suffer if others don't help us. For example, when you see someone having a heart attack on the street, you have a duty to call an ambulance and wait for the ambulance to arrive, even though that involves a bit of effort and missing your bus. Similarly, we might argue that helping the Haves won't cost all that much and that they will miss out on decades of life if we don't, so we have a positive duty to consent to part of our taxes being used to fund life extension research. According to the estimate discussed in section 1.10, Promotion might cost $3 billion a year in research funding—a pittance compared to the entire yearly budget of any sizeable first-world economy.

But I have argued that the Haves are not harmed here. If I'm right, then a positive duty does not arise in that way. However, there's another way to argue for a positive duty here. We might argue that we have positive duties not only to prevent harm but also to confer benefits. Imagine that you have a very wealthy acquaintance who hasn't provided for any family members in his will because he falsely believes they're all dead. He cares about nothing other than his relatives, and because he thinks he has no relatives left, he doesn't care who gets his fortune; his entire estate will go to the Internal Revenue Service. (Assume that his fortune makes no real difference to the federal budget.) You happen to know that his long-lost niece is alive, and if you take the time and trouble to call him up and tell him so, that niece will receive a huge inheritance. The burden of making that phone call is dwarfed by the benefit of inheriting a sizeable fortune, so you have a positive duty to that niece to pick up the phone. Similarly, the Have-nots and Will-nots have a duty to consent to the government spending $3 billion a year on Promotion. Their share of that tax burden is miniscule (as a percentage of their total taxes) compared to what Promotion will eventually provide the Haves.

However, the cost of funding Promotion is not the only burden to consider. If Promotion is successful, then eventually the Have-nots and Will-nots will bear other large burdens in the form of the various harms we've discussed. The burden that's relevant in assessing a positive duty of beneficence is not merely what you have to do for someone else but also what you later suffer or miss out on because you helped. For example, grabbing your hand when you start sliding on the ice may not take much time or effort, but if doing so sends me sliding over the cliff to my death, then that loss is part of the burden. An argument based on the Haves' right to be helped is not promising.

To sum up, making life extension available harms the Have-nots and Will-nots, and failing to make it available does not harm the Haves. Failing to make it available does prevent the Haves from getting a massive gain in welfare. That may or may not be unjust. It may or may not be a serious wrong. However, whatever else it may be, it's not a harm, and so it does not violate their right not be harmed. The Haves' interest in all this is represented not by the right against harm but by the principle that welfare—including theirs—should be maximized. We should not count that interest twice.

Thus, the only rights that are relevant to deciding whether to make life extension available are the Will-nots' and Have-nots' rights not to be harmed. In chapter 11, we will weigh those rights against the welfare gained by making life extension available and ask whether welfare outweighs those rights—or the other way around.

10.6 Conclusions

BB. The right to equality does not weigh in favor of either Inhibition or Promotion, even if we have good reason to believe it will be violated under Promotion. (Section 10.3)

CC. The right to self-determination does not weigh in favor of either Inhibition or Promotion, for neither policy violates the right to self-determination of any of the three groups: Haves, Have-nots, or Will-nots. (Section 10.4)

DD. Promotion will harm the Have-nots and Will-nots, but Inhibition will not harm the Haves. Inhibition prevents the Haves from becoming better off, but that's not a harm. The only right that's relevant to deciding whether to make life extension available is the right of Will-nots and Have-nots against harm. (Section 10.5)

11 Deciding among the Groups—Rights versus Welfare

11.1 Introduction

In chapter 9, I argued that a world with life extension has more net welfare than a world without it. In chapter 10, I argued that maximizing welfare in that way violates the rights of the Have-nots and Will-nots not to be harmed but does not violate any other rights. Maximizing welfare in that way conflicts with respecting their right against harm. How should we resolve that conflict?

In general, we should not maximize welfare at the expense of rights. However, in some circumstances, welfare can outweigh rights. I will propose a test to guide our moral judgment about when those circumstances exist: a gain in welfare is big enough to outweigh rights violations when we would be willing to risk that harm on a chance of getting that benefit if we were making that choice for ourselves. If the welfare of Haves outweighs the rights of Have-nots and Will-nots, then justice requires Promotion. If the welfare of Haves does not outweigh the rights that will be violated under Promotion, then justice requires Inhibition. In this chapter, I'll argue that the welfare of Haves does outweigh the rights of the other two groups, so justice requires Promotion. This is the crux chapter in my long argument in favor of developing life extension.

11.2 How to weigh rights against welfare

As we saw in section 10.2, when rights conflict with maximizing welfare, we usually think we should respect the right even though that means producing less welfare. To reuse an earlier example, we should not kill one person, violating his right against harm, in order to use his organs to save several others who will die without those organs, thereby maximizing welfare (the welfare gained by saving their lives exceeds the welfare lost by ending his life). Ronald Dworkin characterizes this aspect of rights by describing rights as trump cards that can outweigh other moral objectives that are not rights, like maximizing net welfare or making society more equal in some respect, just as a joker trumps any other card in the game of euchre.[1] When increasing welfare or making things more equal can be achieved only by violating some right, then we may not increase welfare or make things more equal in that way or to that extent.

And as noted earlier, rights protect interests. Sometimes the interest in question is the right-holder's own welfare. This means that when a right protecting the holder's welfare conflicts with maximizing the welfare of others, we have a conflict between one person's welfare and the welfare of others. Thus, when a right that protects welfare conflicts with maximizing welfare, what happens is this: the right gives the right-holder's welfare greater weight than the welfare it conflicts with. For example, in the organ case, the healthy person's right against harm gives his welfare greater weight than the combined welfare of the five people his organs can save. If we were weighing welfare loss against welfare gain, then rights that protect welfare interests would be empty—they would protect nothing. When we give someone's welfare the protection of a right, as we do when we say there is a right against harm, we are giving that person's welfare greater weight—stronger protection—that the amount of welfare itself would call for if we were merely weighing net welfare. We are giving that person a trump card.

However, we don't think that we should *always* respect a right when it conflicts with maximizing welfare. There are exceptional cases where most of us would say we should violate some right in order to maximize welfare. Let's revisit the case discussed in section 10.2: Someone has a genetic mutation that makes her white blood cells extremely effective at fighting infections, giving her an abnormally strong immune system—far better than anyone has ever had before. We need to insert a needle to draw a small blood sample so we can clone her white blood cells and develop a serum that gives a similar boost in immunity to everyone else on earth, thereby saving hundreds of millions of lives (now and in the future). She hates needles. She refuses. She reminds us that she has a right against harm. Draw your own conclusions, but my judgment is that she should be forced to give another blood sample, violating her right against harm and her right to self-determination, thereby giving the rest of the human race a spectacular boost in welfare. The harm of being poked with a needle for a couple of minutes is trivial compared to the staggering amount of harm her blood will help us prevent.

So we're not absolutists about rights; we don't think rights trump welfare in *all* circumstances. If the burden the right-holder suffers when his right is violated is sufficiently small and the welfare at stake is sufficiently great, violating that right is the right thing to do. Notice that this is consistent with what we believe about positive duties. A positive duty is a duty to actively do something for someone (rather than merely a duty to refrain from doing something to them). Some philosophers say you have a positive duty when the cost you suffer by helping is sufficiently small compared to the amount of welfare at stake for the person you help (just as welfare might outweigh a right when the welfare is great enough relative to the right).

This is also consistent with what most philosophers say about redistribution to benefit those who are worse off: redistribution from those who are better off to those who are worse off can be justified by improving the condition of those who are worse off even if doing so reduces the total amount of good in that society, but it cannot be justified by merely *marginal* improvements for some at a dramatic cost to those who are better off.[2] For example, many of us might say that justice requires reducing the assets and income of the top third of society by 20 percent in order to improve the lot of the bottom third by 10 percent, but few of us would say that justice requires taking half the top third's wealth in order to improve the bottom third by merely 1 percent.

Because the right against harm gives the right-holder's welfare extra weight, that right can be outweighed by an amount of welfare large enough to exceed the extra, weighted amount of welfare the right protects.[3] To put this another way, we tend to think that morality can require harming someone to get a greater amount of welfare if the gain in welfare is very, very large and the reduction in welfare for the right-holder is much, much smaller.

How do we tell when a given increase in welfare outweighs a right against harm? Philosophers speak of "weighing" or "balancing" duties, rights, and welfare in these situations, but the process of weighing or balancing is not an algorithm—talking this way is just a way of saying that you must use your judgment about which action is right when you must choose between protecting welfare and respecting a right. Consider the scenario where we must decide whether to take a blood sample by force from a woman who hates needles. We consider the situation and reach a judgment about what is right to do. This is what we mean by "weighing."

But what do we really *do* when we engage in such weighing? There is a large philosophical literature on this, but the short answer goes something like this: Gather as much information as you can, make sure you don't have a personal interest in the outcome one way or the other (this isn't always possible, of course), mull things over for a long time, consult with others whose judgment you respect, and then use your best judgment and see which answer "feels" right to you. In the end, that's all you can do.

However, that merely tells us what the experience of forming such judgments is like. We want to know more: What makes such a judgment correct or incorrect? There are two questions here, and it's important not to confuse them. First, there's an epistemic question about how we can know that we've reached the correct judgment. That it "feels" right isn't a guarantee that we came to the right decision. Unfortunately, there is no way to be certain about this. Be as careful as you can when making sure you've thought long and hard, consulted wise people, tried to eliminate your personal interest, and are fully-informed; that's the best you can do.

However, the second question is not an epistemic question about how we know our judgment is correct; it's a metaphysical question about what makes a judgment correct. In other words, it's a question about what kinds of facts or states of affairs a correct judgment detects. Suppose I'm watching a horse race, and I think I see horse A come across the finish line a nose ahead of horse B. My judgment is correct if some part of horse A's body extended farther forward than any part of horse B's body. What plays the role of the horse's nose in judgments about when welfare outweighs a right? One way to approach this question is to ask what rule or principle such judgments seem to fit. It may be that there is some pattern to our judgments about when welfare outweighs a right. If so, perhaps we can state that pattern as a principle. We might be following some principle when we form such judgments even if we're not aware of the principle and can't articulate it. What principle might we be following when we decide whether some benefit to one person or group is large enough to justify some harm to another person or group?

I think this is the principle that best represents the judgments we make in such cases:

The gambling test: A gain in welfare is big enough to outweigh rights violations when we would be willing to risk that harm on a chance of getting that benefit if we were making that choice for ourselves and we didn't know our identity or position in society—and hence didn't know whether we would be a beneficiary or a victim of that harm.

When you don't know your identity or position in society, you're under a partial *veil of ignorance*—a sort of imaginary amnesia about your own identity. Obviously this never happens in real life, but that's all right; the test concerns what you would decide *if* that veil of ignorance were real.

The veil of ignorance falls within a long tradition in ethics, appearing explicitly or implicitly in Hume's "judicious spectator," Kant's categorical imperative, Adam Smith's "impartial spectator," Rousseau's general will, Sidgwick's "point of view of the universe," and Harsanyi's impartial observer. Approaches that use something like a veil of ignorance are said to model sound moral reasoning by suppressing any awareness of one's own interest and of morally irrelevant factors, such as race or gender. That's the argument for treating judgments we make under a veil of ignorance as credible: if we would choose some arrangement under the veil of ignorance, then that arrangement is just, for it's consistent with impartiality concerning morally irrelevant factors like race or gender. The best-known version of a veil of ignorance test comes from John Rawls, who used it to determine which social structures are just. In Rawls's version, people are choosing principles of justice to govern the basic structure of society. They don't know their identity, race or ethnicity, gender, age, talents and endowments, or position in society. Unlike Rawls, the version I am using is a *partial* veil of ignorance, for in the gambling test, people do have some information: they know the percentages of people who will get access to life extension and who will not. In other words, they know the odds of being a Have or a Have-not. However, they don't know whether *they* are Haves or Have-nots, nor do they know which generation they are in.[4]

Notice that the gambling test does not specify the rule of choice it would be rational to use under those conditions. In other words, the test doesn't give you a rule to follow when deciding whether you would take some gamble or other. There are several such rules, and some of them are more conservative than others. At one end of the spectrum of risk aversion, we might follow a rule that tells us to make the choice with the greatest expected utility (e.g., welfare). At the other end of the spectrum, the "maximin" rule tells us to make the choice whose worst possible outcome is better than the worst possible outcome of any other choice. These are only two of many possible rules of choice. One approach to the gambling test is to first decide which rule is the most rational rule of choice under the conditions of the gambling test, with its partial veil of ignorance, and then apply the rule and deduce some decision from it.

However, any argument for a particular rule of choice under the conditions of the gambling test will have to include showing that the rule is consistent with our judgments about what is rational in a given scenario. If, for example, the maximin rule turns out to generate results that don't seem rational to us, then we have reason to think the maximin rule is the wrong rule of choice for that scenario, regardless of a priori arguments that it is the most rational rule of choice. In other words, if the rule conflicts with our judgments about which choices are rational, then we should go with our judgments and set aside that rule. Consistency with such judgments is the best argument for a rule of choice. I believe the correct rule is the one that matches our judgments about when to take the gamble, and to discern that rule, we need those judgments first. Moreover, once we have the judgments, we don't really need to identify the rule. (It would be interesting to do so, but not necessary for our purposes.) Besides, such rules are most useful when we need to make our moral judgments quickly. Big questions like the ones addressed in this chapter deserve more time and reflection. It's better to approach them directly rather than letting a rule guide our judgment.

However, we do need an argument that the gambling test itself is appropriate for determining when a given amount of welfare outweighs some set of rights violations. There are three such arguments. We just saw the first one: the gambling test, like other versions of the impartial spectator/veil of ignorance tradition, ensures the kind of impartiality that is part of justice and models moral reasoning to some extent. Therefore, judgments reached under the veil of ignorance are valid moral judgments.

Second, when we impose a policy that treats one group worse as the price of increasing the welfare of another group, we are treating both groups as if they took a gamble and this was the result. This is just if and only if it is a gamble they *would* take (or would if they were ideally rational, choosing under suitably idealized conditions). This argument for the gambling test is analogous to one way of understanding social contract arguments: we can assess the justice or injustice of a social arrangement by asking whether it is an arrangement people would agree to if they had a choice. Similarly, we can ask whether imposing a policy with different implications for different groups is just by asking whether people would agree to that arrangement before knowing which group they belong to. The gambling test models that choice.

The third argument is that the gambling test fits our judgments about cases where we conclude that welfare has outweighed some rights violation, such as the needle/immune system case—or at least, I hope to show this. If I succeed, then I will have provided further support for the gambling test. I'll try to do this in sections 11.3 and 11.4, where I apply the test to Have-nots and Will-nots.

Now imagine a society where everyone has roughly the same amount and level of welfare as everyone else. Suppose that you're a member of that society, and you're deliberating about a policy that would give 1 percent of the population 5 times as much benefit as they had before but harm the other 99 percent by reducing their welfare by 5 percent per person. (These figures are arbitrary; I use them here just for illustration.) Because you are under a partial veil of ignorance, you know the percentages of the two groups and the amount of harm or benefit they receive, but you don't know your own identity or position in society, so you must assume that your odds of being among those who benefit are equal to their percentage of the population: 1 percent.

Now for the crucial step. To determine whether that increase in welfare outweighs the rights of the other 99 percent not to be harmed, ask this question: Would you (or better yet, some hypothetical rational, reasonable person), under that veil of ignorance, be willing to risk a 99 percent chance of having your own welfare reduced by 5 percent on a 1 percent chance of having your welfare increased five times over? If you would take that gamble, then the net welfare outweighs the right to be harmed in that situation. If you would not, then it does not.

This particular example may be a close call. I probably would not take that gamble, but I can imagine some people taking it. People might disagree about whether the gamble is worthwhile. Of course, we might say that all hypothetical rational, reasonable people will make the same choice, but if they don't, then perhaps this is simply a case where the truth is somewhat vague or indeterminate—we are in a gray area where welfare might or might not outweigh some rights. I won't try to settle that question here; if moral truth is indeterminate in that way, then this is a problem for any moral analysis and not just mine. If it's not, then my discussion is no more troubled by this than any other approach to balancing rights against welfare.

How dramatic a rights violation does this analysis permit? Could we, for example, justify slavery using the gambling test? Let's imagine a case: Suppose that if we enslave 1 human out of every 100, the other 99 people gain a huge increase in welfare. On those odds, people might indeed take that gamble. However, is this a realistic scenario? How would enslaving 1 percent of the population produce such a dramatic gain for the other 99 percent? All we thereby achieve is getting one person's labor for free and making that person do things he or she doesn't want to do. That typically doesn't produce a huge gain in welfare for 99 other people. Slavery doesn't generate dramatic gains in welfare for others unless the percentage of slaves is much higher. However, once the percentage of slaves is much higher, the odds of becoming one are much higher too, and the gamble looks a lot less attractive.

We can imagine outlandish cases where severe, dramatic injustice to one person results in vast amounts of welfare for everyone else, but such cases don't really support the objection that the gambling test is wrong because, under outlandish conditions, it justifies slavery. As Richard Hare argues when he defends utilitarianism against the objection that it justifies slavery, the wrongness of slavery, like the wrongness of anything else, must be shown in the world as it actually is.[5] Our moral principles, including those that express rights, are devised for the kinds of situations and human nature we actually encounter. When we consider outlandish cases like this, we can decide that under those outlandish cases, slavery really would be justified and conclude that it seems wrong only because we are used to more normal cases where it really is wrong. In that event, the slavery cases that pass the gambling test don't support an objection to that test, for the moral objection would not survive in that outlandish world—we merely mistakenly think it would. Consider, after all, the range of rights violations many people already consider acceptable: unintended but foreseeable civilian casualties from military bombing raids in a just war, quarantining people with a lethally contagious disease, or the mere fact that we accept a certain level of risk in product safety. People have a right not to be killed, and we realize that some people will be killed if we allow private automobiles to be used instead of forcing everyone to use mass transit. We tolerate those deaths because we realize that the odds of any one of us being killed in a car crash are pretty low, and the convenience and benefit of having private automobiles is thought to be great enough to justify those deaths. And is death really worse than slavery? Fortunately, plausible real-world cases where enslaving a tiny minority benefits everyone else substantially are extremely scarce, and life extension will not make them more common.

11.3 Weighing Have-not rights against welfare

Now let's use the gambling test to see whether the welfare of Haves is likely to outweigh the rights of Have-nots and Will-nots. We will consider the Have-nots in this section and the Will-nots in the next section. The right in question is the right against harm. As we saw in previous chapters, the possible harms to Have-nots include distress, increased death burdens, Malthusian harms, entrenched elders and opportunity drought, a slower rate of innovation and social change, and challenges to existing arrangements for pensions and elder care. However, as we saw in section 9.5, if life extension is possible at all, it's likely that at some point in the future, the world will be rich enough and the technology cheap enough for everyone to have access to life extension. Presumably that will remain the case from that point on, more or less indefinitely. Thus, in the vast majority of future generations, everyone who wants life extension will be able to get it. In effect, everyone will be a Have. (Again, I'll consider Will-nots in the next section.) Therefore, the Have-nots are a very small percentage of the entire future human race. Finally, under the partial veil of ignorance used in the gambling test, the parties don't know what generation they belong to—that is, where they exist in time. They know that they live in some century anywhere from the advent of life extension into the far, far future, but they don't know when.

So suppose you must choose whether to live in a world where life extension is developed at some time or other, and you don't know which century you'll live in. Because the vast majority of generations in the future will consist of nothing but Haves, there's a very high probability that you'll be a Have. When we ask whether justice permits us to violate the rights of some minority in order to increase or protect the welfare of the majority, it doesn't matter morally whether the two groups exist at the same time. We might, for example, have to decide whether to impose some harm on future generations in order to benefit those living now, or vice versa. Even though the generations don't exist at the same time, the weighing of rights and welfare works the same way. All that matters is the size and importance of the interests involved. In this case, the Have-nots are in the majority in earlier generations but in the minority over the entire future span we are considering. In order to decide whether the welfare of future generations outweighs the interests of Have-nots in the near future, we must consider all those generations, and that requires a thought-experiment where the gambler does not know which generation she belongs to.

For purposes of this thought-experiment, let's suppose that the percentage of Haves in the entire future human race will be 99 percent. Let's further suppose that I was wrong in chapters 2 and 3, and extended life has some serious drawbacks, so that each Have suffers a 25 percent reduction in welfare for each year of extended life, just to give our critics the benefit of the doubt once again. Finally, let's stick with the assumption that each Have-not suffers a 50 percent reduction in welfare. You are choosing whether to risk a 1 percent chance of cutting your lifetime welfare in half in hopes of a 99 percent chance of increasing your lifetime welfare by 8.8 times as much welfare (1,000 years reduced by 25 percent then divided by 85 years). If we consider the entire future history of the human race, the percentage would probably be less than 1 percent, but let's use a 1 percent figure to keep the math simple.

Would a rational agent take that gamble? Most of us would, I suspect.[6] If you're not sure, then consider this case for comparison: Suppose you're suffering from a fatal disease that can be cured with a drug that speeds up the dying process in 1 percent of those who take it but restores normal life span in the rest. In other words, it doesn't lengthen life beyond a normal life span; it restores a normal life span. If you have that disease, would you turn down that drug, thereby forgoing the rest of your life, just to avoid a 1 percent chance of dying from that disease even faster? Very few of us would.

Now alter the case: imagine that the drug does not restore your normal life span, but—as an unavoidable and unintended side effect—*increases* your life span by several times. Would the fact that the drug *lengthens* your life be a reason to be more fearful of that 1 percent chance of speeding up your disease than you are when merely restoring normal life span is the alternative? Surely not. (Quite the opposite, in fact.) You might reject this if you consider extended life so undesirable that given a choice between a life span shorter than normal and a life span longer than normal, you would choose a shorter-than-normal life span. However, if my arguments about the desirability of extended life are sound, very few people would make that choice.

Once we take the long view, it appears that the welfare gained by Haves really does outweigh the rights violations of Have-nots, even if we assume that Forced Choice is not imposed. Of course, if we assume that Forced Choice is imposed, then the Malthusian consequences are prevented and the total harm is substantially reduced. In that case, the gamble looks even better.

Objection: If the invention of life extension is inevitable and Prohibition is impossible, why does it matter whether we follow Promotion or Inhibition? The gambles considered so far involved two alternatives. One alternative was some version of world B, where life extension is developed. The other alternative was world A, where life extension is never developed at all. However, if I'm right and life extension is possible, it will be developed sooner or later. This fact has a very important consequence, for it means that our choice is not between a world with life extension and a world without it but between Promotion, where life extension is developed sooner, and Inhibition, where life extension is developed later.

This fact gives rise to an objection. The objection begins with the fact that there are three time periods to consider. First, there is the period before life extension is developed, a period free of any injustice related to the development of life extension.

Second, there is the period when life extension has recently been developed, and the Have-nots greatly outnumber the Haves. Let's arbitrarily suppose that period lasts for 50 years (it doesn't matter exactly how long it is). Let's also suppose that life extension technology is in its infancy; it confers only 50 extra years of life. Finally, let's suppose that only 5 percent of the human race can get life extension and that the harms caused by making life extension available to that 5 percent reduces everyone else's welfare by 30 percent.

Would you run a 95 percent risk of reducing your welfare by 30 percent in exchange for a 5 percent chance of increasing your welfare by the equivalent of 50 extra years of quality life? Probably not. At best, it's a close call for many of us. We can tweak the figures back and forth, but unless the harm to Have-nots is trivial, this looks like a bad gamble. That harm may be trivial, but I'll give critics of life extension the benefit of the doubt and assume that the harms are significant.

But there's a third period to consider: the vast expanse of time after the first 50 years, when everyone is a Have. Under Promotion, the first period is shorter and the second period is still 50 years, but that 50-year period begins sooner and ends sooner. Under Inhibition, the first period is longer and the second period (the same 50 years) begins later and ends later. All else being equal, the degree of injustice and the number of people affected during the second period is the same under both policies, so neither involves any more injustice than the other. The difference, in short, is that the 50-year period occurs earlier under Promotion and later under Inhibition. (The same is true if we suppose that it takes longer than 50 years to make life extension available to everyone who wants it; the second period is longer, but that makes no difference when we choose between the policies.)

The objection, then, is this: If the development of life extension *is* inevitable, it doesn't matter morally which policy we follow. There's nothing wrong with Inhibition and, in particular, nothing wrong with Inhibition through neglect—the funding policy we have right now. Of course, this doesn't mean that Inhibition is more just than Promotion, but it does mean that Inhibition is no less just than Promotion. The objection shows that it doesn't matter whether we follow Promotion or Inhibition.

However, this objection fails for two reasons. The first reason is that the objection assumes the number of years in which everyone is a Have is infinite. If the human race were going to last forever, then starting later would not result in fewer Haves receiving life extension; it would merely start that infinite series later on. If the third period (when everyone has access to life extension) is infinitely long, then both futures contain the same amount of welfare (an infinite amount). However, it's not likely that the future of the human race is infinite. It's much more likely that our species will come to an end at some point, however many thousands or millions of millennia from now. (Science tells us that eventually the universe will either collapse back into a dense ball resembling the one that turned into the Big Bang or expand forever and suffer a "heat death," where all temperatures fall to an extremely low level. Either way, we can't avoid extinction forever.) If the future of the human race is not infinitely long, then starting life extension sooner provides life extension to Haves that much sooner and thereby provides it to that many more Haves. Even if we can't know for sure that the human race will come to an end one day, it is far more likely that it will and that probability justifies Promotion. Promotion provides extended life to a larger number of Haves, and it does not harm a larger number of Have-nots; therefore, Promotion is more justified than Inhibition.

The second reason the objection fails is that it shows, at most, that there is no reason to prefer Promotion over Inhibition—or vice versa. It doesn't show that there's a reason to prefer Inhibition or to oppose life extension generally. Therefore, even if the first argument (which assumes that the future of the human race is not infinitely long) seems like a very strange argument to rest on, at least it's an argument that favors Promotion over Inhibition. There's no countervailing reason to favor Inhibition.

11.4 Weighing Will-not rights against welfare

I have argued that two long-range trends will gradually make it possible to provide life extension to everyone: the falling cost of life extension and rising levels of prosperity. In time, there may be no Have-nots. However, this is irrelevant to the Will-nots, who have access to life extension but don't want it. Making life extension available to more people will not reduce the percentage of Will-nots over time. Taking the long view may not be relevant here. What does justice require when it comes to the rights of Will-nots?

The answer depends on the variables. Suppose, for example, that half the population are Haves and half are Will-nots. You don't know whether, in the extended world, you're a Have or a Will-not. In other words, you're under a veil of ignorance that prevents you from knowing your values or preferences, so you don't know whether you would want extended life—that's why you don't know whether you're a Have or a Will-not. Suppose further that the harms to Will-nots reduce their welfare by 25 percent.

You're deciding whether to risk a 50 percent chance of suffering a 25 percent reduction in lifetime welfare in exchange for a 50 percent chance of multiplying your lifetime welfare by roughly 12.5 times (1,000 years divided by 80). I'm not sure whether I would take that gamble; it's a close call. I don't have any confident sense of how many people would. If a hypothetical rational person would not take that gamble, then the rights of Will-nots outweigh the welfare gained by the Haves.

Now suppose the harm to Will-nots drops to 10 percent or less and the odds of being a Will-not drop to 10 percent. You're considering whether to take a 10 percent chance of having your net welfare reduced by 10 percent in exchange for a 90 percent change of multiplying your net welfare by 12.5 times. That looks like a much more attractive gamble. If a hypothetical rational person would take that gamble, then the welfare of Haves outweighs the rights of Will-nots.

The problem is that we don't know how much harm the Will-nots will really suffer, though I argued in chapter 4 that it's very little. We also don't know what the actual percentage of Will-nots will be, but the two surveys discussed in section 1.9 provide some clues. The Pew Research Center's Religion and Public Life project surveyed 2,012 American adults about life extension; 56 percent said they would not want life extension if it were available to them.[7] Two Australian universities surveyed 605 adults about life extension; only 35.4 percent said they would take life extension pills if they could.[8] These two surveys suggest that the percentage of Will-nots could be as high as half the population. However, there are two reasons to take those survey results with a grain of salt.

First, the survey respondents knew very little about life extension. In the Pew survey, more than half the respondents had never heard of life extension before taking the survey, 30 percent had heard a little about it, and only 7 percent had heard or read a lot about it. In my teaching and talks about life extension, I find that many people think life extension involves keeping elderly people alive longer, as we do in hospitals now. They often don't understand that life extension involves slowing or halting aging, not end-of-life care of the conventional kind. Perhaps many of the survey respondents mistakenly thought that extending human life means keeping people alive in an infirm, elderly condition far longer than we do now. It may take time for people to adjust to the idea of extended life and become sufficiently informed to form stable views about it.

Second, these surveys revealed some possible inconsistencies in attitudes. In the Pew survey, 68 percent of respondents thought most people would want it, even though slightly more than half of those respondents said *they* didn't want it. The Australian respondents were similar: 65.1 percent said they favored research into life extension technologies even though only a third said they wanted it for themselves. The fact that these attitudes are inconsistent suggests that they're merely proto-attitudes—attitudes that aren't yet carefully considered or stable. They may later shift in one direction or another.

Setting aside qualms about the two surveys, there are two reasons to be skeptical when people currently say they would not want extended life even if it were offered to them. First, it's one thing to declare that you don't want extended life the first time you hear about it, before you've thought about it, and when the prospect is highly remote. It's quite another thing to turn it down when it's a real, concrete possibility. Life extension seems like science fiction right now, and it's hard to seriously consider revising one's sense of the future on the strength of something that speculative. Because the prospect seems so far-fetched, people may affirm their existing attitudes simply because that's easier and because there's no reason to go to the trouble of changing their expectations.

Second, people who say they don't want extended life may be expressing adaptive preferences. (See section 3.9.) An adaptive preference happens when you adapt what you want to what you can get. (I can't reach the grapes, but I didn't want those grapes anyway; they're probably sour.) Throughout human history, we've never expected to avoid aging or live longer than a few decades. We may have adapted our desires about aging and life span to constraints we can't escape. Such adaptation makes the anticipation of death easier to bear. Moreover, it's not plausible that most adults alive today will be able to take advantage of life extension even if it arrives during their lifetime. They're either not young enough or not rich enough—or both. If you have reason to think future generations will get this and you won't, then you have reason to want to believe that it's not worth having. However, future Will-nots are not in that position, and desires concerning death might change when the constraints do.

I'll plant my flag here, for this is the best we can do with the information we now have: So far as Will-nots are concerned, Promotion is more just than Inhibition, because once people become familiar with life extension and have had time to think about it, very few people will turn down extended life if they can get it. The survival instinct is too strong and the aches and pains of aging too severe for any other choice to make sense to most people. (I am leaning, of course, on my arguments in chapters 2 and

3 that extended life is attractive and good to have.) Moreover, if they have doubts, the fact that life extension is a reversible choice makes it easy to experiment with extended life. Anyone who isn't sure can try it for a while and later go off their life extension meds if they change their mind. In light of all this, I'll hazard this prediction: In time, turning down life extension will become as eccentric a choice as choosing lifelong celibacy. It will be an intelligible choice but a rare one.

That line of argument leads to this objection: It seems that in the end, my argument for Promotion rests on a slender and speculative foundation—a set of reasons to think there will not be very many Will-nots. Is that really a sufficient basis for such a momentous policy decision? Perhaps we should hold off until we have more information.

My reply is this: This objection is a sword with two edges, for we also lack sufficient information to be sure that there *will* be lots of Will-nots. Neither Promotion nor alternatives to Promotion have strong factual support, but such support is simply not available until after life extension is available. If we do nothing in hopes of gaining more information, then (a) we will still learn very little until after life extension is available anyway, and (b) we are effectively making a policy choice in favor of Inhibition by neglect. Not choosing a policy *is* a policy choice, and it's no better supported than any other choice.

11.5 Two versions of Promotion

We've been discussing a version of Promotion where the Haves can get life extension as soon as it's available. We might call that *Promotion with Immediate Access*. There's another policy to consider: *Promotion with Delayed Access*, where we develop life extension as quickly as possible, but no one can get it until the technique is affordable enough to be provided to everyone. The rich will just have to wait until society can manage to offer it to the entire population.

Promotion with Delayed Access does not produce as much welfare as Promotion with Immediate Access, since the Haves cannot get life extension as quickly, but unlike Promotion with Immediate Access, it doesn't harm any Have-nots, for it's not available to anyone until the Have-nots can have it too (and thereby become Haves). In short, Promotion with Immediate Access produces more welfare for the Haves at the expense of some rights violations for the Have-nots, while Promotion with Delayed Access avoids those rights violations at the cost of producing less welfare for Haves. Which policy is more just? That depends on whether the increased welfare available under Promotion with Immediate Access is great enough to outweigh the rights of Have-nots. If it is, then Promotion with Immediate Access is more just. If it's not, then Promotion with Delayed Access is more just.

However, I won't explore scenarios where we choose between these two policies, for even if Promotion with Delayed Access is more just, I have two practical concerns that tip my verdict from Promotion with Delayed Access back to Promotion with Immediate Access. First, I don't think we can take Promotion with Delayed Access as a serious option. If life extension is developed, it *will* leak out and get used regardless of attempts to prohibit its use until everyone is rich enough to afford it. Promotion with Delayed Access requires temporary Prohibition, and Prohibition is not feasible. If Promotion with Delayed Access is not possible, it can't be the more just policy—or, at least, attempting to institute it can't be the right thing to do. (You can't have a moral duty to do something you can't do.) Remember, we're doing applied ethics; we want to know what is the right thing to do under the circumstances we find ourselves in, not what is right under ideal circumstances that are highly unlikely to occur.

Second, Promotion with Immediate Access is likely to stimulate the development of life extension faster. The Haves have more money and more political influence; their support will tend to accelerate life extension research, and they're more likely to support it if they think they'll get it. (Look at the money that Google and several Silicon Valley billionaires are pouring into life extension research even now; would they do this if their leaders believed they wouldn't get life extension immediately after its development?) Moreover, the firms that develop it can anticipate bigger and quicker profits if they're allowed to sell it immediately, and that will motivate them too.

Conclusion: so far as the rights of Have-nots are concerned, Promotion with Immediate Access is the most justified life extension research policy.

11.6 Conclusions

EE. Usually, we shouldn't maximize welfare at the expense of rights, but welfare can outweigh rights when the conditions of the gambling test are met. The gambling test says that a gain in welfare is big enough to outweigh rights violations when we would be willing to risk that harm on a chance of getting that benefit if we were making that choice for ourselves and we didn't know our identity or position in society. (Section 11.2)

FF. The gambling test tells us that if we take all future generations into account, the welfare gained by Haves outweighs the rights violations of Have-nots. (Section 11.3)

GG. Even if the development of life extension is inevitable, it matters morally which policy we follow. Promotion is more just than Inhibition even if life extension will eventually be developed under Inhibition too. (Section 11.3)

HH. It's likely that there will be so few Will-nots that the welfare gained by Haves outweighs the rights violations of Will-nots. (Section 11.4)

II. There are two versions of Promotion: Promotion with Delayed Access, where no one gets life extension until it can be made available to everyone, and Promotion with Immediate Access, where the Haves can get it as soon as it is developed. However, even if Promotion with Delayed Access is more just than Promotion with Immediate Access, Promotion with Delayed Access is the morally preferable policy. (Section 11.5)

12   Enhancement Worries

12.1 Introduction

Chapters 9, 10, and 11 were my attempt to pull all the conclusions from previous chapters into a single, all-things-considered conclusion about whether developing life extension is a good thing. I concluded that it is.

In this chapter, I'll discuss a kind of concern that lies outside the line of argument developed in chapters 2 through 11 and requires separate discussion. I am referring to concerns about biomedical enhancements. A biomedical enhancement is any technology or intervention that makes humans physically or psychologically superior in some way—that is, superior to what is normal for human beings. Enhancements might take the form of making humans smarter, or giving them stronger immune systems, or improving their memory, or making them physically stronger, or even improving their moral character by making them more altruistic. These things might be done though drugs, genetic engineering, or some kind of mechanical implant. Enhancements might raise humans to a level that only a few humans currently reach (making us all as smart as Stephen Hawking) or raise them to a level no human has ever reached (making us smarter than any genius in the history of the human race). Life extension is an enhancement if anything is, and when people write about enhancement, life extension is one of the most common examples.

Although enhancement sounds like a good thing, many writers have raised moral concerns about it. Concerns about enhancement first surfaced in public discourse in the late 1990s and early 2000s among a group of bioethics thinkers sometimes known as the bioconservatives. They are not necessarily conservative about all political issues, but they tend to favor conserving the physiological basis of human nature—our genome and all

that rests upon it—largely as it is. They include Leon Kass, Michael Sandel, Francis Fukuyama, and Jürgen Habermas, among others.[1] Sandel and
Fukuyama were members of the President's Council on Bioethics under
George W. Bush, and Kass was chair. Kass began warning against enhancement in the late 1990s, and the President's Council issued two book-length
reports dealing with enhancement in the early 2000s: *Beyond Therapy: Biotechnology and the Pursuit of Happiness* and *Human Cloning and Human Dignity*.[2] The council warned against seeking "perfection" in human nature or
going "beyond therapy."

When bioconservatives warn against enhancement, they're reacting
to a group known as the "transhumanists" or "posthumanists." Transhumanists see human nature, or at least the physical basis for it, as an
engineering project. Nick Bostrom, associated with Humanity+ (the leading transhumanist organization) and director of the Future of Humanity
Institute at Oxford University, defines transhumanism as "the intellectual and cultural movement that affirms the possibility and desirability of
fundamentally improving the human condition through applied reason,
especially by using technology to eliminate aging and greatly enhance
human intellectual, physical, and psychological capacities."[3] Transhumanists advocate improving our intelligence, memory, affect, immune system,
physical strength, and stamina, among other things, using a variety of bio-
technologies, including drugs, bionic implants, nanotechnology, genetic
engineering, xenotransplants, tissue regeneration, cloning, exocortexes
and cyberware (physical devices that mediate between computers and our
brains to improve upon and augment brain function), and even uploading
minds into computers or cyberspace. They are particularly interested in life
extension, but their agenda is far broader than that. That broad agenda
is what the bioconservatives are responding to when they worry about
enhancement and perfecting human nature.

However, although there is some unity to that agenda, no one—including
bioconservatives—argues that enhancement is *always* bad or wrong,
though bioconservatives are often characterized as if they think it is. In
other words, no one thinks all enhancements are necessarily bad. Leon
Kass, who believes the pursuit of "perfection" is the most neglected topic
in bioethics, also says that we should set aside questions about enhancement and ask, "What are the good and bad uses of biotechnical power?"[4]
Kass concedes that perfecting ourselves is not per se wrong: "By his very

nature, man is the animal constantly looking for ways to better his life through artful means and devices; man is the animal with what Rousseau called 'perfectibility.'"[5] It's not fair to charge the bioconservatives with making a blanket objection to all enhancements.

There's a reason no one thinks enhancement is per se wrong: it clearly isn't. Several philosophers have effectively demolished the claim that it is, in case anyone is tempted to say so. Some of them make a continuity argument, pointing out that literacy, running shoes, computers, and many other things enhance human capabilities (to remember, to think, to communicate, to calculate, to run). Those enhancements are not morally problematic; therefore, enhancements in general are not morally problematic.[6] Frances Kamm has a "shifted baseline argument" against the antienhancement view: suppose that what's normal for us were actually abnormal—would it be wrong to improve ourselves to our current baseline? If, for example, our intelligence were naturally 30 IQ points below the current average, would it be wrong to raise them by 30 points? Presumably not. Therefore, there is nothing wrong with enhancement per se.[7] John Harris argues that the distinction between harms and benefits is not morally relevant and that treatments prevent harms while enhancements confer benefits. If it does not matter morally whether you are harmed or fail to receive some benefit, then the treatment/enhancement distinction doesn't matter either. Since treatments are not bad, neither are enhancements.[8]

However, showing that enhancement isn't always bad doesn't show that it's never bad. Various kinds of enhancement or uses of enhancement may be morally problematic for various reasons. There are several concerns that dominate the antienhancement literature. Many of them pertain to some enhancements and not others. I am speaking of "enhancement concerns" in a narrow sense, of course. Broadly speaking, one could say that any ethical issue concerning some kind of enhancement is an enhancement concern; in that sense, this entire book is a book about enhancement ethics. However, such a broad definition of enhancement ethics is not helpful, for there's no ethical issue that is common to all and only biomedical enhancements. We will focus on concerns that are, in some sense, about the very fact that humans are being enhanced, even if none of these concerns assumes that enhancement per se is necessarily bad.

We're concerned with only one kind of enhancement: human life extension. Therefore, rather than trying to assess the entire range of possible enhancements, I will consider the major enhancement concerns one by one, and for each one, I'll ask whether that concern applies to life extension. If it does, I will then ask whether it's a valid concern, at least when it comes to life extension. I will argue that not all enhancement concerns are valid and that none of the valid ones apply to life extension.

As we proceed, bear in mind that life extension is an enhancement in a very particular way. It enhances us by giving humans a life span no human has ever had before, but at any given time during that life span, those humans are lot like humans who have not been enhanced. Assuming they have received no other enhancements, they're not smarter than other humans, or equipped with more powerful memories or immune systems, or physically stronger. They're much like young(ish) humans with normal life spans, except that time and experience probably give them the kind of wisdom that comes with age.

12.2 Risk and the precautionary principle

The first concern we'll consider is not unique to enhancements, but it does arise in a particularly strong form for certain enhancement techniques: nanotechnology, germline genetic engineering (genetic changes that would be passed on to offspring), transplanting genes from other species, and devices (still to come) that would link our brains to computers for added brain power. The human body is a delicate and complicated set of systems that evolved over millions of years, and some writers argue that we shouldn't try to improve upon something so complicated, so important to our welfare, and that we don't fully understand unless we have very, very good evidence that the alteration is not too risky.

Some people may believe that life extension is too risky. We discussed several possible dangers in previous chapters. Some say that extended life might not be good for us. It's likely that making life extension available to the Haves will cause distress and other harms to the Have-nots, possible harms to the Will-nots, possible Malthusian consequences, and other undesirable social consequences. Each of these may either fail to materialize or turn out to be outweighed by the advantages of making life extension available, but still, there's a risk that the downside will outweigh the upside. Developing life extension is something of a gamble.

This is a valid point, but the fact that something is a gamble doesn't mean we should never take that gamble. We must assess the risk and then decide whether the possible benefits outweigh the possible dangers. I'm not talking about assessing what the risks are—that is, estimating the probability of some harm. I'm talking about deciding whether to run some risk once we know the odds—or at least know as much as we can about them.

One way of making such decisions is particularly prominent in discussions of novel technologies: the precautionary principle. There are different forms of this principle, but in general, they all recommend taking action to prevent harm even when the evidence that harm will occur is less than conclusive. There are two classic versions of the precautionary principle, both developed for environmental protection but sometimes proposed for other contexts. The earliest appears in the Rio Declaration on Environment and Regulation (1992), which states:

In order to protect the environment, the precautionary approach shall be widely applied by States according to their capabilities. Where there are threats of serious or irreversible damage, lack of scientific certainty shall not be used as a reason for postponing cost-effective measures to prevent environmental degradation.[9]

The other is from the Wingspread Conference on the Precautionary Principle in 1998:

When an activity raises threats of harm to human health or the environment precautionary measures should be taken even if some cause and effect relationships are not fully established scientifically. . . . In this context the proponent of an activity, rather than the public, should bear the burden of proof.[10]

The Rio version of the precautionary principle is weaker, for it says merely that a lack of "scientific certainty" that serious or irreversible harm may occur is not a reason to postpone measures to prevent that harm. The Wingspread version says that too but also says that protective measures "should" be taken and that the party who wants to undertake the potentially harmful activity has the burden of proving that it's safe enough.[11] When it comes to new technologies, the precautionary principle seems to require us to err on the side of playing it safe and to place much greater weight on the possible risks than on the possible benefits of that technology.

However, while the precautionary principle may be a plausible policy or rule of choice for climate change and some other environmental issues, it's not a good policy for all new technologies. It's highly risk averse. It overlooks the possible benefits of a new technology and fails to consider the possibility that sometimes not intervening carries extreme risks.[12] In other words, the harm some new technology will help us avoid may be greater than the harm that technology might cause. Consider a simple example: Suppose some new drug might cause serious liver damage but has a high probability of eliminating some form of deadly cancer. To prevent harm to the liver, we would have to avoid using that drug and run the risk of dying from an otherwise untreatable cancer. Harris and Holm call this the "precautionary paradox": we are to pause lest we cause harm, but pausing may cause great harm too.[13] In short, the general problem with the principle is that it's often not clear in which direction caution lies—that is, what poses the greater risk of harm: doing something novel to solve a problem or not solving that problem.[14]

So which choice is more risky: normal aging with painful senescence toward the end and death within a century or life extension with its harms? If we're talking about harms to those who extend their own lives, the safer choice is to extend your life and run the risk of boredom and reduced death benefits, especially given that you can always go off your life extension meds (so to speak) and resume aging. If we're talking about harms to the Have-nots (or to everyone), such as Malthusian harms, then we're not talking about risk—we're talking about justice. We talked about justice at length in several earlier chapters. If you're a Have-not or Will-not, the issue is not whether life extension has more upside for you than downside; it has no upside for you at all. The issue is whether it's fair to impose some downside on you so that the Haves can enjoy the upside. That's a serious issue, but it's an issue of justice.

I have always maintained that the Malthusian and justice issues are the most serious objections to life extension, and I don't take them lightly. All I'm saying here is that they aren't issues about decisions concerning risk. The precautionary principle is the wrong instrument for thinking about issues of justice, for to think about justice, we must think about how the harms and benefits are distributed. The precautionary principle takes the standpoint of someone who doesn't know how things will turn out for them. To use it for an entire society obscures the fact that we know quite a lot about how things may turn out for each group and that they won't turn out the same for everyone.

So we don't have sufficient reason to think that making life extension available is the riskier choice. However, there's an interpretation of the precautionary principle that requires no such evidence. Instead, it requires merely that we don't know the degree of risk. Stephen Gardiner argues that the precautionary principle is a version of the maximin rule of rational choice. The *maximin rule* tells us to make the choice whose worst possible outcome is better than the worst possible outcome of any other choice. If we interpret the precautionary principle in this way, it seems to tell us to avoid developing life extension, for the worst possible (though not necessarily likely) outcome of a world with life extension is a Malthusian catastrophe that makes life intolerably bad, while the worst possible outcome of a world without life extension, all else being equal, is that everyone lives a normal life span under conditions where life is not intolerably bad.

However, the maximin rule is meant for choice under uncertainty, not for all choices under risk. A decision about risk is what you make when you know the probabilities of some harm occurring; a decision under uncertainty is what you make when you don't know that probability. Moreover, the maximin rule is not plausible under all conditions of uncertainty. It's plausible when you attach far more value to avoiding the potential harm than you attach to gaining the potential benefit.[15] You might, for example, think that enhancements have something like diminishing returns, that normal health is good enough, and that for anything beyond "good enough," the risks are too high. In other words, you might think that if you already enjoy normal health, the additional gain in welfare promised by some enhancement is less valuable than the same gain in welfare when you are unhealthy and some medical treatment will restore you to normal function and capacity.[16]

So situations where the maximin version of the precautionary principle applies have two features: we don't know the odds, and we attach much more value to avoiding the potential harm than to gaining the potential benefit. Neither of those features is present in the context of life extension.

First, we do have some sense of the odds, at least for some of the possible bad consequences. We know that it's likely that those who lack access may feel distress and sense that their deaths are somehow worse, and we know that it's likely that not everyone will have access, at least for some time. However, we also have reason to believe that someday, it will be possible to provide it to anyone who wants it. We know that there's a non-negligible chance of some Malthusian consequences down the road and that there's some danger the gap between Haves and Have-nots will get wider, perhaps increasing the economic and political power of those who are already better off. We also know that it's possible—though difficult—to control our population well enough to avoid Malthusian consequences and that societies sometimes succeed in equalizing the distribution of wealth and power. And we know that they often fail to do so. Whether the odds and magnitude of these harms are such that we should try to inhibit the development of life extension is debatable, but we know enough to have some sense of the odds involved in that decision. There's some uncertainty about those odds in the sense that we aren't certain what those odds are, but this is not a decision under uncertainty in the strict sense. We do have a rough sense of those odds. This is a decision about risk, so the maximin version of the precautionary principle does not apply here.

The second feature is not present here either, at least not for everyone: not everyone attaches more value to avoiding the risks of life extension than they do to enjoying the benefits of extended life. Many people already say they would prefer to run those risks (or consider them negligible) and enjoy dramatically longer life. Of course, that raises a further issue about how to balance the wishes of those who want to live in a world where that is an option and the wishes of those who want to live in a world where no one has that option, but that too is an issue of justice, not an issue of decisions about risk. For both of these reasons, the maximin version of the precautionary principle is not appropriate for deciding whether to develop life extension and therefore isn't a reason not to develop it.

12.3 Authenticity

One of the most common concerns about enhancement is that it will undermine authenticity. The ideal of authenticity tells us that we should be true to ourselves. According to Erik Parens, the "moral ideal of authenticity . . . is that each of us finds our own way of being in the world. It is my job as a human being to find my way of flourishing, of being true to myself."[17] Authenticity concerns about enhancement take two forms.

First, some writers have objected to certain enhancements, such as drugs that give you greater courage or greater physical prowess, on the grounds that getting such qualities through drugs and not through hard work means those qualities are not really yours. Kass believes that such enhancements pose the risk of "disrupting the relation between what we do and what we become."[18] More generally, if our qualities are installed in us through biomedical enhancements, we either don't develop character or we can't take credit for our character. To develop character, we must struggle to develop other qualities and not get them from a pill or an implant. This concern often arises in connection with the use of performance-enhancing drugs in sports. The resulting athletic achievements are said not to be authentic; the athlete is not fully responsible for those results.

Second, Carl Elliott has argued that Prozac, which can be seen a happiness enhancer that gives you a better personality, also gives you a false self, and a false self is clearly not an authentic self.[19] I suppose this depends on the patient and the drug. Having used antidepressants myself, I can tell you that I feel that I am much more my "true self" on antidepressants than off, but perhaps not all cases are like this. In any case, Elliott's concern is not that we can't take credit for the personality the drug gives us (though perhaps we can't, if all we do is take a pill every morning). His concern is that the personality a drug gives us is not our real personality. However, this concern would arise even if we had to work very hard for a long time to get that drug and *could* thereby take a kind of credit for the result.

I'm not rejecting these concerns, but neither version of the authenticity objection applies to life extension. Consider Kass's version first. Life extension keeps you physically youthful longer than normal, but it doesn't give you courage or strength or anything else you didn't have during your healthy adult years—it just helps you keep that health for a longer time. Whatever you do with that extra time might result in changes to your character, but those are changes you can take credit for. True, you can't take credit for getting the extra years (or maybe you can, if you earned the money to pay for life extension), but you can't take credit for a normal life span either, so that's irrelevant. Moreover, no young person can take credit for being youthful. Life extension merely extends that state of affairs over a longer span of time.

As for Elliott's worry about acquiring a false personality, that too doesn't apply to life extension. Slowing or halting aging doesn't alter your personality in the way that taking Prozac is said to do. If living longer in a youthful state does result in changes to your personality over time, those changes occur in the way that personality changes usually do: over a long time, influenced by the life you lead. If that results in a false self, then we all acquired false selves long ago.

12.4 Sandel's concern about "giftedness"

Michael Sandel is concerned about two kinds of human enhancement in particular: enhancing athletes so that they perform better and enhancing children. He worries that enhancing human beings in these ways exhibits a defect in our attitude toward our own strengths and abilities. His concern is similar to the concern about authenticity, but his concern is not that we won't be able to take credit for our characteristics. Instead, his concern is that we'll take credit for our characteristics when we shouldn't:

> The problem is not the drift to mechanism but the drive to mastery. And what the drive to mastery misses and may even destroy is an appreciation of the gifted character of human powers and achievements.
>
> To acknowledge the giftedness of life is to recognize that our talents and powers are not wholly our own doing, nor even fully ours, despite the efforts we expend to develop and to exercise them. It is also to recognize that not everything in the world is open to any use we may desire or devise. An appreciation of the giftedness of life constrains the Promethean project and conducts to a certain humility. It is, in part, a religious sensibility. But its resonance reaches beyond religion.[20]

What does it mean to say that the "drive to mastery," the project of enhancing people, exhibits a lack of "appreciation of the giftedness of life"? Part of what Sandel means is this: when you appreciate the "giftedness" of life, you recognize that your talents and powers are not wholly your own. You can't take credit for being smart, or good at music, or having a robust immune system and rarely getting ill—you're simply lucky to be that way, and you should be grateful, even humble about it. You fail to appreciate the "giftedness" of life in this way when you think those abilities, strengths, or capacities are something you can take credit for—that they're something like personal achievements.

Moreover, according to Sandel, enhancing ourselves can lead to two further problems. First, we will become responsible for a wider range of traits. For example, parents will become responsible for aspects of their children that are now a matter of genetic luck. Second, we'll have a diminished sense of solidarity with those who are less fortunate, presumably because they aren't able to enhance themselves. This might, for example, undermine the insurance market.[21]

The first thing to say here is that it's not clear that Sandel's concern extends to life extension. Sandel is concerned about designer babies and bionic athletes. He's concerned with ways of giving us advantages or desirable characteristics that we weren't born with. Life extension, however, doesn't give us enhanced intelligence, happier affect, a better memory, or the ability to run a four-minute mile. It does gives us more robust health because it makes us more youthful and the young tend to be robust, but it doesn't give us new desirable characteristics or get rid of any old ones. It merely lets us keep some of our desirable characteristics for a longer span of time.

But let's suppose for discussion that life extension does make us somewhat arrogant about the advantages of youth (as some of us were when we were young). Sandel is right that we should be humble about talents, abilities, and capacities that we are born with or at least acquire through luck. However, he takes this point too far. There are four problems with his argument not just with respect to life extension but with respect to all enhancements.

First, there's no reason to appreciate something as a gift when it's not a gift at all. Right now, having a good memory is a gift (we can think of it as a fortunate endowment even if there's no God). If, however, you pay a clinic to improve your memory, then your memory (or at least the margin by which it was improved) is not a gift—it's something you paid for. Of course, if you received the money as a gift or earned it partly as the result of some competitive advantage in life, *then* perhaps you should regard your improved memory as a gift or at least acknowledge that you purchased it with a gift.

Second, even if we have a duty to be grateful for what is given to us, that doesn't mean we must never improve the gift or that we must be grateful for all improvements to it. Gratitude for whatever intelligence I have is not a reason to refrain from enhancing my intelligence with drugs; at most, it means that I should be grateful for the original quotient of intelligence I started with. It doesn't mean I should be grateful for the increased quotient of intelligence I get from the drugs. More important, it doesn't mean that I should not enhance my intelligence at all, lest I seem ungrateful or unappreciative of the original quotient of intelligence I started with.

Third, it makes no sense to be grateful for *everything* (malaria, reality TV, cockroaches). Here is where Sandel's arguments, even if valid for the enhancements he discusses, break down when it comes to life extension: Why is having an 80-year life span rather than a much longer life span a thing to be grateful for? Refraining from extending life so that we won't seem ungrateful for the gift of that life span is like refraining from chemotherapy so that we won't seem ungrateful that the incidence of cancer isn't higher.

Sandel doesn't mention life extension, but perhaps he would say that extended life is not good for us; therefore, we should be grateful we don't have it. If that's what he means, speaking of gratitude, giftedness, and mastery is an obscure way to say so. Instead, he should talk about whether extended life is a good life to have. I argued in chapters 2 and 3 that extended life *is* good for us, but even if I'm wrong, that problem with life extension is not that we would exhibit a lack of appreciation for our gifts. Rather, that problem would be that we don't know what's good for us. This is a mistake about what's in our interests, not a case of taking personal credit for something that's really a product of luck.

Finally, the value of an enhancement can outweigh the moral weight of whatever duty of gratitude we might have; the importance of gratitude does not always outweigh everything else. I should be grateful for a coat I received for Christmas, and it may seem ungrateful to exchange it for something else at the store, but if the coat doesn't fit me, I should take it to the store and exchange it for one that does. The value of my health and comfort outweighs the duty to be grateful.

However, Sandel denies that other moral considerations might outweigh the duty or virtue of gratitude. His argument is somewhat obscure, so let's see the passage:

I am suggesting [that] . . . the moral stakes in the enhancement debate are not fully captured by the familiar categories of autonomy and rights, on the one hand, and the calculation of costs and benefits, on the other. My concern with enhancement is not as individual vice but as habit of mind and way of being.[22]

Here he says that what's wrong with lacking appreciation for giftedness is not a matter of good or bad consequences, or rights and self-determination, or the virtues. For most philosophers, those three categories exhaust the range of things that can have moral value, but Sandel believes there is more:

The bigger stakes are of two kinds. One involves the fate of human goods embodied in important social practices—norms of unconditional love and an openness to the unbidden, in the case of parenting; the celebration of natural talents and gifts in athletic and artistic endeavors; humility in the face of privilege, and a willingness to share the fruits of good fortune through institutions of social solidarity. The other involves our orientation to the world that we inhabit, and the kind of freedom to which we aspire. . . . Changing our nature to fit the world . . . deadens the impulse to social and political improvement.[23]

This passage raises two issues.

First, norms of unconditional love for one's children, humility, and a willingness to share good fortune with those less fortunate seem to fall within familiar moral categories. They might be seen as virtues (love, solidarity, generosity, humility), or as duties (to love unconditionally, to help those worse off), or as consequences that are good (children who are loved unconditionally are better off, and sharing good fortune tends to shift resources to those who will receive more utility from them that those who already have enough). In any case, they don't seem to fall outside "the familiar categories of autonomy and rights, on the one hand, and the calculation of costs and benefits, on the other."

Second, even if Sandel has identified a new moral category, he hasn't established that this new moral category, whatever it is, always takes moral priority over anything else it might conflict with. The mere fact that it does not involve appeals to consequences, virtues, or duties, does not entail that it's morally more important than consequences, virtues, or duties. If he thinks it's morally more important than any other moral considerations, he needs to explain why. Merely showing that it's different (if he's done even that much) doesn't show that it can't be outweighed by other moral considerations.

12.5 It's not natural

The enhancement literature contains many references to nature, human nature, and something called "the natural." Some observers take that to mean that opponents of enhancement think that what's natural about human nature has moral value by virtue of being natural and that altering it results in a less natural, and hence less valuable, human nature. The first thing to say about this is that few (if any) bioconservatives or other opponents of enhancement have made such an argument. However, it appears that some people are concerned about this, so let's see what might lie behind this concern.

Richard Norman attempts to reconstruct an argument for this concern.[24] Norman doesn't believe that interfering with nature is necessarily wrong, nor does he find this argument convincing, but he believes it might help us understand this concern better. The argument doesn't require claiming that altering what is natural in human nature is per se wrong. Instead, it raises a concern about *hyperagency*—"a state of affairs in which virtually every constitutive aspect of agency (beliefs, desires, moods, dispositions and so forth) is subject to our control and understanding."[25] In other words, we can choose to control any aspect of human nature that influences what we choose to control and how. (This appears to be a close cousin to Sandel's concern about giftedness.)

According to this argument, hyperagency would undermine the meaning of our lives. The argument rests on two claims: a "logical" claim and a psychological claim. The logical claim is that there's no standpoint from which to regard one choice as better than another if *everything* is a matter of choice. At least some things have to be good or bad, right or wrong, independently of our choices and preferences, or we can't make choices in a nonarbitrary way. In order to make "meaningful" choices, we need things that are beyond choice. He calls these "background conditions."[26]

The psychological claim is that humans are disposed to use features of their biology as background conditions. These conditions tend to be facts about birth and death, maturing and aging, having to work in the world and overcome adversity, or sexual relations and relations among the generations. Exactly where these boundaries are drawn varies from culture to culture and generation to generation. For example, some people object to using ovarian tissue from aborted fetuses in assisted reproduction or to using Prozac and other antidepressants on the grounds that these interventions alter aspects of our biology (and perhaps psychology) that must not be altered.

Norman contends that if we alter these aspects of the human condition too much, we are threatened with a loss of meaning in our lives. Those who find contraception unnatural, for example, may believe that sex is more meaningful when it's closely tied to reproduction and that we lose some of the meaning in our lives when we separate sex and reproduction. Those who feel this way have identified the procreative aspect of sex as a background condition. Those who object to antidepressants as unnatural may feel that part of life's meaning comes from struggling against adversity, grappling with pain, and that happiness without a struggle lessens that meaning.[27] Those who feel this way have identified our unmodified range of moods as a background condition.

People who object to enhancements as unnatural may not put their objection in these terms, but Norman believes that something like this lies behind their concerns even if they don't realize it. As I said, Norman is trying to reconstruct a possible argument for a point of view he doesn't really share. He believes the logical claim is true, but he also believes that which aspects of our biology seem beyond choice is something that changes over time and varies from culture to culture and that objections to altering human nature are usually not strong objections, for we can incrementally adjust our sense of what is natural. Over time, we can decide, for example, that using ovarian tissue or Prozac is acceptable. The boundaries of what is natural migrate over time. Presumably we then find meaning in life in other ways. For example, using ovarian tissue may cease to threaten the meaning of our lives when we find the meaning of childbirth and parenting less in genetic continuity with the parents and more in parenting and nurturing itself.[28]

I have no quarrel with Norman's psychological claim as a description of a common way of thinking, though many people (including myself) do not think that way. That said, I think the argument works (if at all) only on an implausible reading of the logical claim. The logical claim is correct insofar as we have no basis for choosing one thing over another unless "something is accepted as given and not open to choice." We can't make moral choices unless we have normative standards, principles, or values that enable us to rank our options, and those normative standards cannot be yet another set of options equally open to choice. However, neither the logical claim nor the psychological claim, nor both together, imply that the "given" is some feature of the natural world, let alone some feature of human biology. They imply merely that *something* has to be the standard and that many people *think* the standard has something to do with not altering what is natural. The argument Norman presents does not establish that those people are right—that what is natural operates as a moral standard. Moreover, if people or societies have some discretion over which aspects of human biology are a background condition, then a dilemma arises. Either the background condition is a matter of choice, which contradicts the logical claim, or at least some of those people or societies mistakenly identified something as a background condition when it's not, in which case we should be all the more skeptical of what people think is the standard.

The argument for concern about what is natural seems fatally defective, but before we move on, let's look at Norman's suggested way of applying it to life extension. He suggests that our background conditions include "the temporal and biological structure of a normal human life."[29] Our natural life span, in other words, can be seen as a background condition, and if we extend our life span too far, we have defied the background condition and our lives will lose some of their meaning. (It's not clear whether Norman agrees with this application of the argument or uses it merely for illustration.) Many of our projects, and their meaning, assume the scale of a normal life span, and their importance depends partly on their location in that life span (the book you want to finish before you die because it's the culmination of your life's work, for example). Norman agrees that our life span could get longer without losing those aspects but believes that living forever (or at least for millions of years) would deprive us of them.[30]

Notice, however, that this argument can be made without any reference to "the natural" as something it would be wrong to alter. All this argument requires is the claim that events and experiences get some of their meaning from their relation to other parts of our lives and their place in the overarching narrative of a life span. We saw that argument in section 3.10, and it didn't require any reference to the natural. We saw a very similar argument in sections 2.2 through 2.4, where we considered the concern that extended life might become intolerably boring, and another similar argument in section 3.6, where we considered the argument that an extended life might have less meaning because valuable things become less valuable the more of them we have, and the longer we live, the more we get. None of these arguments require the claim that a natural human life span is something we cannot alter too much because it's natural. They require merely the claim that if we alter our life span, it may lose meaning of a certain kind. The fact that our current life span is natural is beside the point.

12.6 The value of a natural life span

Let's consider another concern about a natural life span. Extended life is longer than our natural life span and hence unnatural in that respect. Is an unnaturally long life span morally problematic?

Proponents of such an argument would have to explain what has moral value: the fact that our life span of roughly 80 years is natural or the fact that it's a life span of 80 years? Neither version makes much sense so far as our interests are concerned. The former version suggests that any life span that is natural is best for us, so if we had a natural life span of 45 years, then a 45-year life span would be best for us, which is silly. The latter version of the objection suggests that having a natural life span of 125 years would somehow be worse for us, which is equally silly.

Moreover, it's arguable that our current life expectancy is not truly a natural life span anyway. As I discuss at greater length in section 12.9 and section A.4, there are three mainstream theories of the evolution of aging. (It's possible that all three of them are correct to some degree.) All three rest on the fact that even if we never aged, we would live only so long in the environment where we evolved. Consider this example: Suppose you had a mouse that never aged. How long would that mouse live in its natural habitat? According to field biologists, about two and a half years, for the number and size of predators, the incidence of disease, and the availability of food in its native habitat are such that death is highly likely by the time two and a half years have passed even if the mouse doesn't age. It takes metabolic energy to defend against the cellular-level processes of aging, and if the mouse isn't going to last longer than two and a half years anyway, it makes more sense to allocate that energy to feeding, reproduction, fighting, and fleeing. That is just what evolution has selected, for as it happens, roughly two and a half years is a mouse's natural life span. Evolution did not select for longer-lived mice because, in their natural habitat, mice would not live longer anyway. Why waste metabolic energy making them last longer?

The same is true for us. We evolved at a time when we had no agriculture and had to deal with large predators (the Pleistocene era, saber-toothed tigers). The archeological record tells us that we died by our mid-40s. This means that our natural life span is around 45 or so; living into our 70s is an unnatural effect of civilization, and our current life span is unnatural.

Moreover, our natural life span is a function of our size relative to other predators and a few other factors during the Pleistocene era. It does not result from an evolutionary design process that worked to satisfy our interests as we individual humans see them. There is no reason to attach moral value to a natural life span and no presumption in favor of what evolution hath wrought. The fact that nature gave us a particular life span is not a reason to avoid seeking to extend it.

12.7 Playing God

Sometimes those who introduce new biomedical technologies are accused of "playing God." This way of putting things is not common among bioconservative writers, so we shouldn't lay this at their doorstep. However, it is common among laypersons who express opposition to some forms of biotechnology, so let's address it.

Playing God can mean a couple of things. First, we might play God in the sense of trying to do a job that's reserved for God—invading his jurisdiction, as it were. What that means depends on your theology, but one might argue, for example, that God has a plan for the natural order and that humans are allowed to modify that design only within certain limits: we can create new breeds of livestock but not new breeds of human.

This version of the objection requires an explanation of what is and what is not in God's jurisdiction and why. Even among believers, there's no consensus about what jurisdiction God reserved to himself when it comes to biotechnology or what his plan calls for. Moreover, some citizens who hold religious beliefs also believe that we shouldn't appeal to such beliefs when we make arguments to one another about matters of public policy. Finally, some citizens consider all religious beliefs false. These problems don't mean that arguments about divine jurisdiction are wrong, but they do indicate that disputes about this are likely to be interminable. Those who wish to make such an argument are welcome to take up the challenge, but merely making the accusation of "playing God" is not enough to meet that challenge.

Second, there is a nontheological, cautionary sense of "playing God." This means taking on a task we don't have the information, expertise, or wisdom to do well. The idea is not that we're violating divine law, just that we're taking on a task we're not smart enough to do well. This way of warning us against playing God is another way of warning us to be careful, go slow, and not overestimate our powers. This is a theistic way of raising concerns about risk—a concern we considered when we discussed the precautionary principle.

12.8 Fukuyama, human nature, and human rights

Francis Fukuyama is a bioconservative who claims that human rights are based on human nature. He says that if we alter human nature, we endanger the rights that rest upon it: "Human reason . . . is pervaded by emotions, and its functioning is in fact facilitated by the latter. Moral choice cannot exist without reason, needless to say, but it is also grounded in feelings such as pride, anger, shame, and sympathy."[31] Fukuyama believes that if we alter human nature, we'll lose our moral status (presumably by losing our capacity for making moral judgment) and thereby lose our moral rights.

It's not clear why losing one's capacity to make sound moral judgments results in losing one's moral status or moral rights. Animals, after all, have moral status: it's wrong to torture them. Perhaps Fukuyama means that we can't have the full range of rights we do have unless we have a faculty of reason, for only creatures that are rational and self-aware have the kind of moral status we do—and the kind of rights that go with that moral status. It's plausible that there are some things we can't reason about unless we also have a capacity for emotion and affect generally and that morality is one of those things. However, that doesn't amount to losing our ability to reason altogether. Moreover, even if we change our affects and emotional responses, we're not eliminating them—we merely have different emotional responses and affects than we used to have. (No one is talking about turning humans into Vulcans.) Without knowing more about the changes in question, we can't even say whether those changes degrade the kinds of reasoning that Fukuyama has in mind—for all we know, those changes might improve them.

In any case, even if Fukuyama's objection is sound with respect to enhancements that alter our affective responses, his objection does not apply to life extension. There's no reason to expect that slowing or halting aging or living for centuries will disrupt the relationship between affect and thinking, or degrade our capacity for moral judgment, or strip away our affect and capacity for emotional response. If anything, living far longer might improve our moral judgment, as life experience tends to do.[32]

12.9 Is aging a disease?

Occasionally you hear a suggestion that life extension is wrong because aging is not a disease. As with some of the other objections considered in this chapter, it's hard to find a bioconservative (or anyone else) who actually says this. It does come up now and then, however, and it's closely related to the objection that life extension is wrong because it's an enhancement, for enhancements are often distinguished from treatments, and treatments treat diseases while enhancements (it is sometimes argued) do not treat diseases. The connection, then, is that if aging is not a disease, then slowing aging must be an enhancement. Thus, arguing that aging is not a disease is a way of arguing that life extension is an enhancement. Of course, that point was conceded at the outset of this chapter for reasons unrelated to whether aging is a disease.

Still, the treatment/enhancement distinction does pertain to a particular kind of objection to enhancement, for that distinction might be relevant when we're deciding what society should pay for in a program of subsidized universal healthcare. The treatment/enhancement distinction is related to the concept of "medical necessity" in health insurance. One can argue that paying for conditions that aren't diseases or impairments may result in paying for extravagant personal preferences, such as cosmetic surgery for people who are merely plain or growth hormones for children who are merely slightly shorter than average. One can also argue that we have a moral right to healthcare based on a right to equal opportunity—we need to achieve normal function in order to have the normal range of opportunities but above-average function would give us above-average opportunity.[33] However, arguments that health insurance and subsidized healthcare need not cover enhancements are largely irrelevant to life extension. I argued in earlier chapters that life extension should be subsidized for the Have-nots, but those arguments are grounded in egalitarian arguments that have nothing to do with healthcare and did not assume that everyone has a right of access to life extension because aging is a disease.

Aside from concerns about health insurance and equal opportunity, it's hard to understand what's wrong with "fixing" conditions that are not diseases. Just because something is not a disease doesn't mean we have no reason to prevent or remove it, and even if life extension isn't medicine, that doesn't mean we shouldn't practice it (as if only the only legitimate way to make our physical lives better is medicine).[34]

That said, this issue does raise some interesting questions in the philosophy of biology concerning how to define aging and disease, and I will close this book by discussing those questions. I will conclude that aging is a disease.

The best thinking about how to define aging isn't done by philosophers, it's done by biologists, so let's see what they have to say. Some biologists deny that aging is a disease on the grounds that aging is universal to all members of a species, while diseases are not.[35] Some go further and say that unlike aging, a disease can sometimes regress or be cured or treated, though not all diseases can regress on their own. (Of course, this distinction breaks down if we learn how to slow or halt aging.) Aging is not a disease, according to Leonard Hayflick, because

biological aging is a loss of molecular fidelity—an increase in molecular disorder throughout the body caused by random events—that occurs after reproductive maturation in animals that reach a fixed size in adulthood. Biological aging is an expression of the Second Law of Thermodynamics, or increasing entropy, or disorder, in a system. Aging is not a disease.[36]

Hayflick is referring to the fact that most organisms (including us) produce, order, and replace molecules with fidelity up to around the time the organism is likely to reproduce, given the hazards of that organism's natural environment. Natural selection does not give organisms the ability to maintain molecular fidelity indefinitely, perhaps because organisms that devote energy to repairing and maintaining molecular integrity are thereby devoting less energy to feeding, breeding, fleeing, or fighting and thus may not reproduce as much as they do.[37] Hayflick sees this as entropy at the cellular level, and entropy is not a disease. We can put his point this way: Entropy is no more a disease than getting hit by a car while crossing the street is a disease. Both processes disrupt the body's molecular organization. The latter process is merely messier and faster.

Other biologists argue that aging *is* a disease. David Gems, a geroscientist at University College, London, argues that aging is a multifactor genetic disease, and the diseases of aging are symptoms of an underlying syndrome. According to Gems, aging is a disease in spite of being universal. It involves dysfunction and deterioration at the molecular, cellular, and physiological levels, and the diseases of aging are symptoms of a larger syndrome: aging itself.[38] In a similar vein, philosopher Arthur Caplan has argued that aging can be regarded as a disease. The fact that aging is inevitable and universal does not mean aging is not a disease—so are high blood pressure, sore throats, colds, and tooth decay, yet they are diseases. He also argues that aging results from a lack of evolutionary foresight; it has no function or purpose for us. Finally, the processes of aging look a lot like disease processes: they produce discomfort, they have symptoms and manifestations, and they impair our functions.[39]

The debate presented in the previous two paragraphs is happening at too high an altitude. To settle that debate, we need to get closer to the biological ground and then ask whether we draw a valid distinction between aging and age-related diseases. There are three possibilities. First, aging might be nothing but a collection of diseases that tend to correlate with age. In that event, aging is not distinguishable from disease at all. It then follows that aging itself is a disease or at least a set of diseases. Second, aging might include age-related diseases—that is, those diseases are part of aging (which diseases are part of it may vary from organism to organism)— but also includes something else that's not an age-related disease. In other words, aging might be a combination of age-related diseases and something else that's not a disease. If aging includes some element that is not a disease, then it may turn out that we cannot fully treat the age-related diseases without also slowing or halting aging. In that case, treating age-related diseases may not be completely distinguishable from slowing or halting aging itself. In other words, the distinction between medicine and life extension might break down even if the distinction between aging and age-related diseases still holds, at least in part. Third, aging might not include age-related diseases at all.

So are aging and age-related diseases completely distinct? If they are, then it's possible to age without disease and die of causes other than disease or violence. Geroscientists who think they are completely distinct sometimes speak of "normal aging": aging without disease. Normal aging happens when someone ages without any heart disease, cancer, Alzheimer's, diabetes, Parkinson's disease, or any other maladies that are associated with age and also considered to be diseases. The concept of normal aging has a counterpart in the concept of "natural death": death without disease. Those who die a natural death die of old age and nothing else. Everyone agrees that normal aging and natural death—if they exist at all—are quite rare.[40] However, not everyone agrees that there *are* such things as "normal aging" and "normal death."

It's also possible that there are some age-related diseases that are completely distinct from aging itself and other age-related diseases that are not.[41] (This is the second possibility mentioned above.) For example, some geroscientists distinguish between diseases whose pathogenesis seems to involve basic aging processes (such as type 2 diabetes, osteoporosis, cerebrovascular disease, Alzheimer's disease, and Parkinson's disease) and diseases that don't involve basic aging processes but are age-related, either because they occur more frequently in the aged (such as gout, multiple sclerosis, amyotrophic lateral sclerosis, and many cancers) or because they have more serious consequences in the aged (such as some infectious diseases). Those in the first group would be aspects of aging itself, while those in the second and third groups would be age-related diseases.

It's often hard in practice to distinguish between aging and age-related disease. Consider some examples. Bone loss is age-associated, but it's considered a disease only if it has the magnitude to have a clinical impact.[42] Until 1990, systolic blood pressure of 140–160 mmHg was not considered hypertension; now it is, and counts as a disease.[43] Atheroma (lesions of atherosclerosis) can be considered a degenerative process if it results from molecular-level or cellular-level changes in the artery wall, but it can be considered an age-related disease if it results from injury or infection of the vascular wall.[44] Cataracts (arguably an age-related disease) result from normal changes in proteins that increase their opacity (arguably part of aging). Menopause is age-related, but it poses an increased risk of osteoporosis and atherosclerosis. The prostate gland enlarges as men age and leads to hormone changes that can result in cancer.[45]

Another reason it's hard to distinguish between aging and age-related diseases is that aging makes organisms more vulnerable to stress and infections and increases the incidence and severity of accidents and disease. Infectious diseases, for example, are often worse or more common in the elderly.[46] For that reason, one could argue that a particular episode of some infectious disease is an age-related disease on the grounds that the patient would not have been infected at all if he were young. One could even hold that people do not die from age-related diseases; they die from whatever changes in the body made them more vulnerable to such diseases.[47] Finally, when people are elderly, often there are so many age-related diseases that it's hard to single out any of them as a cause of death. When that happens, perhaps we should say that the patient died from the underlying cause of all those diseases: aging itself.

I conclude that we can't distinguish, either conceptually or in practice, between aging and age-related diseases, and therefore, aging is a disease—or at least a set of diseases, where the members of the set vary from organism to organism.

Life extension is a continuation of medicine by other means.

12.10 Conclusions

JJ. Life extension is an enhancement. (Section 12.1)

KK. There is nothing wrong with enhancement per se, and no one thinks there is. (Section 12.1 and Sections 12.3 through 12.8)

LL. The precautionary principle may or may not be appropriate for the environmental context, but it's not appropriate for all new technologies. It overlooks the possible benefits of new technologies and fails to consider the possibility that not intervening might be more risky. Normal aging poses a greater risk of harm than extended life. (Section 12.2)

MM. The maximin principle does not apply to the choice between normal aging and extended life, for the odds are not completely unknown and it's not the case that we attach far more value to avoiding the potential harm life extension might bring than we do to gaining its potential benefits. (Section 12.2)

NN. When we apply the precautionary principle or the maximin principle to the choice between the funding policies of Promotion and Inhibition (not the choice between normal aging and extended life), we're no longer dealing with a decision about risk. We are dealing with questions of justice. (Section 12.2)

OO. Objections to enhancements on the grounds that they undermine authenticity, that they involve failing to appreciate giftedness, that enhancements are not natural, that we should defer to evolution as a designer, that they involve playing God, or that they undermine the basis for human rights are either not good objections to begin with or don't apply to life extension. (Sections 12.3 through 12.8)

PP. Aging is a disease, and life extension is a kind of medicine. (Section 12.9)

13 Policy Recommendations and List of Conclusions

In this chapter, I summarize the policy recommendations I've argued for here and there in various chapters and then list all the conclusions reached in chapters 2 through 12. The list of conclusions serves as a narrative index for the course of argument developed in this book.

13.1 Policy recommendations

I have addressed three policy issues: (1) whether developing life extension is a good or bad thing (that is, whether it should be promoted or inhibited), (2) how to handle the fact that not everyone can afford access to life extension, and (3) how to minimize the possibility of Malthusian population pressures.

Policy concerning the development of life extension
The first policy choice is whether to encourage or discourage the development of life extension. In chapter 1, I presented three possible public policies concerning the development of life extension:

Inhibition: We can try to inhibit such research—prevent it from going forward or at least slow it down. We might refuse to spend public funds on it, or fund it only very conservatively, or try temporarily to prohibit it until we develop a complete regulatory framework or a strong social consensus about such research.

Prohibition: We can try to prohibit anyone from using life extension once it has been developed.

Promotion: We can fund life extension research aggressively in order to develop life extension methods as quickly as possible.

Throughout this book, I have argued for the policy of Promotion. In particular, I argued (in section 11.5) for Promotion with Immediate Access (Haves can get life extension as soon as it is developed rather than waiting until all the Have-nots can get it too) rather than Promotion with Delayed Access (no one can get life extension until everyone who wants it can have it). There are two reasons I favor Promotion with Immediate Access. First, Promotion with Delayed Access is a form of temporary Prohibition, and Prohibition is not feasible. Second, Promotion with Immediate Access will probably stimulate the development of life extension faster, making it available to earlier generations of Have-nots as well as earlier generations of Haves.

Promotion is presented as a policy concerning public funding for scientific research and whatever laws might further or inhibit life extension research, but the idea is more general than that. This is a way of saying that life extension is, on balance, a good thing, and we are better off getting it sooner rather than later. We could say that without reference to public funding for research, of course, but deciding how to spend our tax dollars is a natural place to raise the question. Even if Silicon Valley multibillionaires fund the research all by themselves and bring life extension about before anyone else has had time to mull it over, we should still ask whether that is a good thing. (We might, after all, want to think about whether to encourage or discourage them from doing so.) Deliberating about these three policies is a way of doing that.

Policy concerning access to life extension

The second policy choice concerns access to life extension once it has been developed. In chapter 8, I argued for two sets of policies:

Who gets access

• No one should be denied access to life extension, nor should the development of life extension be inhibited. (Section 8.3)

• Life extension should be subsidized for those Have-nots who want it to the extent possible, provided that the amount of wealth necessary to do this does not exceed the amount that justice requires to be transferred from those who are better off. (I leave open what that amount is, but I believe that once the technology matures, it won't be too difficult to provide it to all of them, so I have not tried to address this.) (Section 8.5)

• If the amount needed to subsidize it for all Have-nots is larger than what justice requires, then instead of subsidizing it for some Have-nots and not others (selected, for example, by lottery), justice requires equal compensation to all Have-nots in some form other than subsidized access to life extension. This might be done through cash transfers or other social benefits. (Section 8.6)

How to finance access for the Have-nots
• To the extent that subsidizing life extension for the Have-nots requires an increase in the overall transfer of wealth that justice would require in the absence of life extension, that increased transfer should come solely from the Haves. This additional burden can be financed at least in part through a user fee or use tax on Haves or an increased income tax above an age past which people don't live without life extension (such as age 115, or perhaps a graduated increase from 90 on up), perhaps coupled with a confiscation of much of their assets. (Section 8.5)
• Will-nots should not be subject to greater transfers of wealth than justice would require in the absence of life extension. (Section 8.5)

Policy concerning the possibility of Malthusian consequences
The third policy choice concerns the threat of increased population pressures that will occur (in the absence of any limits on reproduction) if people live far longer than they do now. There is some room for debate over how much population increase constitutes a Malthusian crisis. I suggested in section 6.2 that an increase of one-third is tolerable. I might be wrong. The main point is that we can set a target fertility rate to keep population growth within whatever limit we decide upon.

In sections 6.3 through 6.7, I argued for a reproductive policy I call "Forced Choice," which restricts the number of children people can have after they use life extension but does not apply to those who never use life extension. Because Forced Choice doesn't apply to those who don't use life extension, it doesn't guarantee that we won't exceed our target anyway. If we do, then we may have to impose reproductive limits on everyone, or set the limit for those who use life extension even lower, or some combination of the two.

Here are the elements of Forced Choice (for illustration, I'll assume a limit of 0.5 children per woman for however long that is necessary):

The reproductive limit and who it applies to

• Those who extend their lives can have (in our example) an average of only 0.5 children per woman at an average age of 25 at childbirth.

• This policy is imposed on everyone, regardless of gender or sexual orientation.

• It is applied on an individual basis, not on couples as units.

How we decide who gets to reproduce if the limit is low enough that not everyone can

• To decide who gets the right to reproduce, those who wish to extend their lives must enter a reproductive lottery or be denied access to life extension.

• If you've had no children yet and you enter the lottery, you have a 25 percent chance of winning the right to reproduce.

• If you've *already* had one child and you then enter the lottery, the trade-off is reversed: you enter the lottery with a 25 percent chance of winning the right to extend your life. Thus, having one child reduces your chances of being allowed to extend your life from 100 percent to 25 percent, and if you win the lottery, you may not reproduce again.

• If you've had *two* children already, then you're forbidden to have life extension.

• Delaying the average age of childbirth slows the population increase to some extent, so your odds of winning the right to reproduce are correspondingly higher the longer you wait to enter the lottery (linked to whatever higher rate of childbirth at that average age of childbirth will produce the same population increase).

• If you lose the lottery when seeking to reproduce at a given age 25, you can subtract your odds at that age from your higher odds at some later age and enter the lottery again at that later age with a percentile chance of reproducing equivalent to the difference between the original odds and whatever higher odds you would have had if you had waited until that later age to enter the lottery the first time. This can be repeated indefinitely, with progressively lower odds of winning.

How you can use your right to reproduce

• If you win the right to reproduce, anyone you reproduce with can do so as well even if they lost the lottery.

• Those who extend their lives may have children with people who don't extend their lives, but the reproductive limit still applies to the person who lives an extended life.

• The reproductive limit does not apply to adoption.

• If you're living an extended life and your child dies, you can have another child to replace the one who died. The reproductive limit is based on the number of children, not the number of childbirths.

• Someone who is biologically unable to reproduce still gets a 25 percent chance to reproduce. If that person wins the lottery, they can then use assisted reproductive technology and egg and/or sperm donors (or even cloning) to exercise that right and have a child.

• People who win the lottery can sell or give the right to reproduce to someone else, somewhat like a carbon credit.

What if you extended your life and you then reproduce without the right to do so?

• If you've already extended your life and you then have a child without having won the reproductive right to do so, your life-extending treatments are terminated and you resume aging at a normal pace. Your reproduction is now no longer limited (assuming your society has not imposed some other limit on reproduction that applies to everyone), but you will resume aging.

What if you extend your life and you later change your mind?

• If you extend your life and you later wish you had reproduced instead (or wish you had more children than you were allowed to), you can reverse that decision and revert to normal aging, just as if you had never extended your life in the first place. You can then have more children.

• If lots of people extend their lives and then, after many decades or centuries, they decide to discontinue their life-extending treatments and *then* have an extra child, the reproductive limit could be reimposed on that group so that they don't increase the population too much (due to the lingering guest problem).

How to enforce Forced Choice

• Governments must keep track of birth records to see who has had children and how many children each person has had.

• Life extension clinics would be licensed and supervised, and each person who comes in would be identified (perhaps with a pinprick blood sample and a DNA chip) and their reproductive records called up to see how many children they have had.

• Genetic testing (again, with a pinprick of blood and a DNA chip) would be used to read every child's genome. The child's genome would be compared with all others in the system so the child's biological parents (or at least sperm or egg donors) can be identified. There would be no penalty for the child, but if a child is traced to parents who have exceeded their reproductive limit, then the parents are penalized.

• Birth records would be checked every day; anyone above a certain age is obviously extending their life. We would take a pinprick blood sample from anyone above the age of 115 who isn't already in the database of people permitted to use life extension and use a DNA chip to identify that person's genome. That genome can then be compared with the genetic records for all children, and if any of the children turn out to be a match with that adult, we know that adult has reproduced and how often. We then check to see how many he or she has already had, and so on.

It's easy to object that these methods of enforcement are not airtight, and of course, they are not. However, like most other laws and policies, Forced Choice need not be 100 percent effective in order to prevent a Malthusian crisis. Being enforced well enough is good enough. Besides, a Malthusian crisis would develop over a long period of time, with plenty of opportunity to fine-tune the policy and the methods of enforcement.

13.2 List of conclusions

I will finish by listing, once again, all the conclusions reached in each chapter, partly so that it is easier to see how we got to the policies discussed above, partly to sketch the overall argument in this book, and partly so that readers can locate particular issues more easily:

A. Extended life *could* become very boring, but it probably doesn't have to be. You might avoid boredom by acquiring new and repeatable interests and projects over time. (Section 2.3) You might also avoid boredom by taking a pill, much as we now use pills to avoid depression, anxiety, and other states we do not like. There's nothing wrong with taking boredom pills. (Section 2.4)

B. Even if you change so much over time that eventually you become a different person, either because you acquired new interests and values or because you just lived long enough, this is not a reason to turn down life extension. It's rational to have an *ongoing desire*, a desire at any given time to continue existing a while longer. Moreover, evolving into a new person is at least no worse than descending into senility and moving closer to death, and probably better. (Section 2.5)

C. Even if extended life is unavoidably boring, that's not a good reason to refuse extended life. At most, it's a reason to discontinue extended life after it has become unavoidably boring. (Section 2.6)

D. There are reasons to believe that extended life would be just as good as life in a normal life span, all things considered. (Section 3.2)

E. The bioconservative arguments tend to exhibit four mistakes: they use arguments that work only (if at all) for immortal life, they overlook counterintuitive implications that arise when we extend those argument to all forms of immortal life, they cannot be run backward without absurd results, and they look only at the possible downside to extended life. (Section 3.3)

F. There is insufficient reason to believe that extending the human life span will make it harder to accept death, undermine our motivation and make us waste time, be less meaningful than life as we know it, or make it harder to develop virtues and care about things beyond our selves. (Sections 3.4 through 3.8)

G. Extended life presents a new human condition, especially in radically extended lives. The new human condition has four features: (1) aging is elective, (2) life extension is reversible, (3) death will be unscheduled, (4) your life expectancy at any given age will always the be same, and death will deprive you of vastly more time no matter how old you are when you die. The new human condition might be better than the old one. (Section 3.10)

H. To the extent that there are benefits to living a normal life span, refusing life extension but living in a world where life extension is available might reduce death benefits for the Will-nots to some degree. However, any harm of this kind is likely to be very minor. (Section 4.2)

I. Refusing or discontinuing life extension is a kind of suicide, but not an immoral suicide. (Sections 4.3 and 4.4)

J. Will-nots who live in a world where life extension is available and who believe that refusing or discontinuing life extension is an immoral suicide may feel that they've been forced to choose between an extended life they don't want and a form of suicide that is morally wrong. Will-nots who believe this suffer a harm called "moral injury" (though not all Will-nots have that belief about life extension and suicide, so not all of them suffer a moral injury). (Section 4.4)

K. Most of the concerns about undesirable social consequences, such as pension-funding concerns or worries about hierarchies becoming more entrenched, are not trivial but probably not insurmountable. Some of them may well turn out not to be problems at all. In any case, living in a world with these problems is not worse than death. Moreover, the human race has adapted to other, equally profound changes in its history. It's time to do so again. (Section 5.2)

L. There is reason to be optimistic that a world full of people living extended lives will be in many ways a much better world: more mature, more interesting, stabler, wiser, more concerned about the future and the environment, and more at peace. (Section 5.3)

M. There is good reason to believe that making life extension widely available will pose a serious risk of a Malthusian crisis of overpopulation, pollution, and resource shortages. To avoid this, it is very likely that we will have to significantly limit the reproduction of those who extend their lives. (Section 6.2)

N. A policy of Forced Choice, which requires people to choose between extending their lives and having as many children as they wish, is feasible and just. (Sections 6.3, 6.4, 6.5, 6.6, and 6.7)

O. The distress suffered by Have-nots over the fact that others can get life extension and they can't is a harm, even if they have nothing to be distressed about. We may have a duty to alleviate it to some extent. However, it should be discounted unless there is another harm they are distressed about. Their distress is a separate harm from that other harm. (Section 7.2)

P. Making life extension available harms the Have-nots by increasing their death burden. The measure of the severity of death—of the death burden—is counterfactual C: How long you would have lived with life extension, discounted by the odds of getting life extension (in other words, if you would have lived another 1,000 years but your odds of getting life extension were only 10 percent, then your loss is equivalent to 100 years). (Section 7.3)

Q. Inhibiting or prohibiting the development and distribution of life extension can reduce the death burden, but it can never eliminate it. (Section 7.4)

R. A world where the Haves get life extension and the Have-nots don't is unequal in ways that are unjust. (Section 8.2)

S. Once life extension has been available to Haves for long enough, it may become reasonable for people to think that extended life is, so to speak, normal—that having enough in life includes having extended life and that a normal life span is no longer sufficient. (Section 8.2)

T. Equality doesn't require leveling-down, so the fact that life extension will not be available to everyone is not a justification for retarding the development of life extension (Inhibition) or preventing people from having it (Prohibition). (Section 8.3)

U. We should not postpone or prohibit life extension until other, more pressing needs are met first. For example, the fact that some people lack basic healthcare is not a good reason to prevent life extension from being developed or distributed. (Section 8.4)

V. Even if society should redistribute resources from wealthier members to poorer members, the advent of life extension does not increase the size of transfer that justice requires from Will-nots, though it may require reallocating that transfer. However, it does require increasing the size of transfer required from the Haves, perhaps by imposing a user fee or increased tax on those who extend their lives. (Section 8.5)

W. If it's not possible to provide life extension to all Have-nots who want it, justice is not served by providing it to some Have-nots but not others. Instead, justice requires that Have-nots should be compensated in some way that can be distributed equally among them, such as a redistribution of wealth or increased social spending that benefits all of them. (Section 8.6)

X. Even if we can be sure that many Haves will evade their duty to subsidize or otherwise compensate Have-nots, Prohibition and Inhibition are not justified. (Section 8.7)

Y. The best way to decide which life extension funding policy is morally most justified is to apply a set of midlevel moral principles that call for maximizing welfare whenever doing so either does not violate a right or violates a right but the amount of welfare is very, very large compared to the importance of whatever interest the right protects without violating any rights. (Section 9.2)

Z. In the long run—and quite possibly in the short run too—a world with life extension will almost certainly have greater welfare than a world without it. (Sections 9.3 through 9.6)

AA. It's very likely that a world with life extension will have a larger population than a world without it and have greater welfare than the world without it. (Section 9.7)

BB. The right to equality does not weigh in favor of either Inhibition or Promotion, even if we have good reason to believe it will be violated under Promotion. (Section 10.3)

CC. The right to self-determination does not weigh in favor of either Inhibition or Promotion, for neither policy violates the right to self-determination of any of the three groups: Haves, Have-nots, or Will-nots. (Section 10.4)

DD. Promotion will harm the Have-nots and Will-nots, but Inhibition will not harm the Haves. Inhibition prevents the Haves from becoming better off, but that's not a harm. The only right that's relevant to deciding whether to make life extension available is the right of Will-nots and Have-nots against harm. (Section 10.5)

EE. Usually, we shouldn't maximize welfare at the expense of rights, but welfare can outweigh rights when the conditions of the gambling test are met. The gambling test says that a gain in welfare is big enough to outweigh rights violations when we would be willing to risk that harm on a chance of getting that benefit if we were making that choice for ourselves and we didn't know our identity or position in society. (Section 11.2)

FF. The gambling test tells us that if we take all future generations into account, the welfare gained by Haves outweighs the rights violations of Have-nots. (Section 11.3)

GG. Even if the development of life extension is inevitable, it matters morally which policy we follow. Promotion is more just than Inhibition even if life extension will eventually be developed under Inhibition too. (Section 11.3)

HH. It's likely that there will be so few Will-nots that the welfare gained by Haves outweighs the rights violations of Will-nots. (Section 11.4)

II. There are two versions of Promotion: Promotion with Delayed Access, where no one gets life extension until it can be made available to everyone, and Promotion with Immediate Access, where the Haves can get it as soon as it is developed. However, even if Promotion with Delayed Access is more just than Promotion with Immediate Access, Promotion with Delayed Access is the morally preferable policy. (Section 11.5)

JJ. Life extension is an enhancement. (Section 12.1)

KK. There is nothing wrong with enhancement per se, and no one thinks there is. (Section 12.1 and Sections 12.3 through 12.8)

LL. The precautionary principle may or may not be appropriate for the environmental context, but it's not appropriate for all new technologies. It overlooks the possible benefits of new technologies and fails to consider the possibility that not intervening might be more risky. Normal aging poses a greater risk of harm than extended life. (Section 12.2)

MM. The maximin principle does not apply to the choice between normal aging and extended life, for the odds are not completely unknown and it's not the case that we attach far more value to avoiding the potential harm life extension might bring than we do to gaining its potential benefits. (Section 12.2)

NN. When we apply the precautionary principle or the maximin principle to the choice between the funding policies of Promotion and Inhibition (not the choice between normal aging and extended life), we're no longer dealing with a decision about risk. We are dealing with questions of justice. (Section 12.2)

OO. Objections to enhancements on the grounds that they undermine authenticity, that they involve failing to appreciate giftedness, that enhancements are not natural, that we should defer to evolution as a designer, that they involve playing God, or that they undermine the basis for human rights are either not good objections to begin with or don't apply to life extension. (Sections 12.3 through 12.8)

PP. Aging is a disease, and life extension is a kind of medicine. (Section 12.9)

Appendix A

The Science behind Life Extension

In chapter 1, we covered some of the science behind life extension, particularly possible methods of life extension, the basic processes of aging, and how long we might live. That was enough to get the book going, and I didn't want to delay the discussion of ethics. However, some readers may want to know more about the science behind all this. There is good reason for that: knowing more will deepen your understanding of life extension, and the science is fascinating. This appendix does not repeat the material presented in chapter 1; it complements that material.

A.1 Defining aging

I defined life extension as slowing, halting, or reversing some or all aspects of aging, but what is aging itself? First, consider several things that aging is *not*. Aging is not merely the passage of time: two organisms might be the same chronological age, but one might have aged more than the other (it looks and feels older). Moreover, aging is not merely the visible physical changes we associate with old age, such as gray hair, wrinkles, fading hearing and failing vision, or loss of bone tissue, for different species and different individuals may display different outward signs of aging even when their life spans are the same and they have aged the same amount. Nor should aging be identified with whatever cellular- and molecular-level processes lie behind increasing frailty over time, for it's conceivable that two species both exhibit aging but have different cellular- and molecular-level processes of aging. We have a clear intuitive sense of aging (we know it when we see it), but for a more precise definition, we need to consider how geroscientists define aging. Their definitions tend to fall into two groups.

The universality definition

Some geroscientists define aging in terms of universality: unlike disease, aging is found in *all* members of a species. Bernard Strehler, for example, says that aging has four characteristics: it is universal (found in all members of the species), intrinsic (inherent to the organism and not a response to environmental factors), progressive (it gets worse over time), and deleterious. Some geroscientists add two more characteristics: aging is also irreversible and genetically modified.[1] The universality element is the heart of these definitions, however, for some diseases are intrinsic, progressive, deleterious, irreversible, and genetically modified, but none of them is universal.

The risk of failure definition

Most geroscientists define aging in terms of increasing risk of failure.[2] According to this definition, aging is simply a state of affairs where, over time, organs and other systems in the body are increasingly likely to fail. According to the risk of failure definition, the very fact that they're more likely to fail at 30 than they were at 20, for example, *is* aging. This definition began with Benjamin Gompertz, the English actuary who discovered the "Gompertz curve": after age 30, the odds of dying double every 7 to 10 years (depending on the country). John Maynard Smith has this in mind when he defines aging as "a progressive, generalized impairment of function resulting in an increasing probability of death."[3] The increasing probability of death is due to failures in function that result in sickness, disease, frailty, and/or reduced speed or endurance. Does this mean that all failures of biological function are part of aging? If my appendix fails at age 25, have I thereby aged? Presumably not; the failures might be *associated* with aging (glaucoma, for example), but aging itself is whatever makes the odds of such failure greater over time, even if we think of the failures themselves as part of aging.

Advantages of the risk of failure definition

The universality and risk of failure definitions might be coextensive in what they pick out; it may turn out that whatever causes increasing risk of failure in a species is also universal to all members of a species, and vice versa. That said, I favor the risk of failure definition, for it has some operational virtues. First, we can use it to date the beginning of aging in an individual: an individual starts to age when the risk of failure begins to increase. (Steven Austad suggests that human aging begins at the age where the odds of death are at a minimum: age 11.[4]) Second, we can also use it to measure the rate of aging—just measure the rate at which failure is increasing. Third, the risk of failure definition enables us to entertain the possibility that some aspects or processes of aging might not be universal—that is, the possibility that some individuals might suffer increasing risk of failure in certain processes or systems and others might not. (My intuition is that this is aging even if it's not universal.) Finally, the risk of failure definition doesn't require a sharp distinction between aging and age-related disease. We do want to distinguish diseases that aren't part of aging, but that doesn't require claiming (as universality definitions do) that *no* age-related diseases are part of aging; the risk of failure definition allows us to entertain that possibility too. I am not claiming that they are part of aging, only that we should not settle the question by defining them out of the concept.

So life extension is slowing, halting, or reversing increases in the risk of failure in an organism.

A.2 Previous gains in life expectancy

The first thing to understand about life extension is that it will be nothing like the gains in life expectancy we've seen over the last 150 years. Those gains didn't come from slowing aging. They came from eliminating things that kill us before we reach our natural life span.

The maximum natural human life span is roughly 120 years. It hasn't changed in 100,000 years, but few of us live much past 80. It could be worse; the archeological record indicates that for most of human history, hardly anyone lived past 45. Life expectancies increased partly because societies achieved a more stable food supply through agriculture, partly because of improved sanitation and living conditions, and partly through eliminating infectious and parasitic diseases. As a result, life expectancy at birth in developed countries is now in the upper 70s or higher. People often think this means that adult life spans are longer than they were a few centuries ago, but that's a mistake; between 1860 and 1960, life expectancy at birth in the United States increased by 31 years, but the life expectancy for an adult who reached the age of 60 increased by only 1 year during that century. In other words, a century ago, those who made it to 60 lived roughly as long as we do, but the odds of making it to 60 were much worse. A high percentage of the increase in life expectancy comes from reducing infant and child mortality. Note that none of this involved slowing the rate at which people age.

This increase in life expectancy is known as the *epidemiological transition*. Thanks to this, people in developed countries rarely suffer or die from infectious diseases. Instead, they suffer and die from age-related chronic and degenerative diseases, such as cancer, heart disease, stroke, and diabetes. However, unlike what happened when we eliminated infectious diseases, eliminating these diseases would not produce big gains in life expectancy. Eliminating all cancer, heart disease, stroke, and diabetes would increase our average life span from 82 years to 96 years—an increase of 17 percent.[5]

We're nearing the end of what can be done to extend the human life span by eliminating disease. From here on, the only way to dramatically increase life expectancy is to slow the rate at which we age.

A.3 Aging is puzzling

We're so familiar with aging that we take it for granted. However, aging is very puzzling for several reasons. First, it's a puzzle from the standpoint of evolutionary theory. The longer an organism lives, the more offspring it can have. Natural selection favors organisms that produce more offspring. Given that organisms *can* age more slowly (as the experiments mentioned above indicate), we would expect evolution to select for organisms that age very slowly or not at all and therefore have more offspring. However, evolution has not done this.

Second, organisms develop from single-cell fertilized eggs into vastly complex living things. Creating a complex organism is more difficult than maintaining one. If creating an organism is harder than maintaining it, we should expect that organisms would maintain themselves indefinitely and never age. After all, they've already performed the more difficult task. However, they don't.

Third, geroscientists have recently discovered that we gradually cease to age in our mid-90s.[6] Aging is an increase in the risk of failure for an organism or its components; the older you are, the more likely you are to have a heart attack, for example. Our risk of failure starts to increase during late youth and gets worse as we get older. However, it levels off around age 95. This is called the *late-life mortality plateau*: if you live to around 95, you remain aged and grow older, but you do not *age* any further. Of course, your risk of failure is so high by then that you eventually die of some age-related cause. Even so, our bodies naturally stop aging if we just live long enough.

Fourth, aging is puzzling because some organisms *never* age, or they do it too slowly for us to notice. Rockfish, the bristlecone pine, lobsters, sturgeons, the microscopic hydra, and some sharks do not age at all—accidents, disease, and predators finish them off. Male flounders age but female flounders don't (they just keep getting bigger). Some colonies of corals are more than 20,000 years old.[7] Bacteria don't age; they divide into daughter cells—an ending that's not quite a death. Other species age but nonetheless regenerate limbs and other parts of their bodies; starfish, some reptiles, and all plants can do this. Even we do this to some extent, for most cells in the human body are less than 10 years old and most will be replaced several times during your life. Within each cell, molecules are constantly being broken down and replaced, even in cells that are not themselves replaced, thereby rebuilding the cell from the inside out. Our germline cells—the cells that produce eggs and sperm—don't age at all.

A.4 Why we age

So why *do* we age? Aging is puzzling from the standpoint of evolutionary theory, and the answer to that puzzle lies in evolutionary theory. There are four evolutionary theories of aging that claim to solve that puzzle.

The mutational accumulation theory

The first evolutionary explanation of aging is the *mutational accumulation theory* associated with Peter Medawar.[8] Some genetic defects appear early in the life of an organism and others appear late in its life. This theory builds on the fact that if an organism has a genetic defect that appears early in its life, the defect will make the organism vulnerable in early life, before it can reproduce, or at least before it reproduces as much as it's capable of doing. If that happens, the genetic defect is less likely to be passed on to offspring—or at least not to very many of them—for the organism will die before it has very many offspring. Thus, natural selection works against genetic defects that appear early in life.

However, things work out differently with genetic defects that appear later in life. A defect that appears later in life is likely to be passed on to a larger number of offspring, for the organism will reproduce before the defect weakens it. If the defect appears late enough, the organism is likely to die from predators, disease, or resource shortages before the defect materializes. Thus, defects that don't appear until that time or later are unlikely to be weeded out by natural selection, for those defects don't affect reproductive success. Finally, the later in life the defect appears, the less likely it is that natural selection will weed it out of the population. For example, progeria (a genetic disease that mimics aging and kills by the time of adolescence) occurs only through random mutation—it's never inherited—because its victims die before they can reproduce. (Because it's random, it's also quite rare.) However, Huntington's disease doesn't kill its victims until they're in their 40s and have had a chance to reproduce. Natural selection doesn't weed this defect out of populations, and it therefore runs in families.

The result is that harmful genetic mutations that occur only late in life will accumulate—hence the name "mutational accumulation theory." According to this theory, the phenotypic changes we call aging are caused by genetic defects that did not, before the rise of agriculture and civilization, affect individual survival, for humans died of something else first. Therefore, those genetic defects were passed on to offspring. However, under civilized conditions, most people live long enough for those genetic defects to manifest themselves, and thus we age.[9]

However, George C. Williams noticed some problems with this theory. First, animals in the wild do sometimes die from senescence. They may not appear old, but senescence is gradual in most species, and even a bit of aging may slow an animal down enough for a predator to catch up. Another problem is that some of the genes involved in senescence are not random mutations but have been around for millions of years; baker's yeast, nematode worms, fruit flies, and mice all share some genes involved in aging. These species do not all die at around the same age in their natural habitat, so the age when they are statistically likely to die from predators or disease does not obviously explain how they all came to share these genes.

The antagonistic pleiotropy theory

Williams proposed the second major evolutionary theory of aging: the *antagonistic pleiotropy theory*.[10] Genes exhibit "antagonistic pleiotropy" when they have both beneficial and harmful effects. Williams surmised that some genes might have a beneficial effect early in life and a harmful effect later in life. According to his theory, the genes involved in aging have beneficial effects early in life, while late-life effects result in the changes we call aging.[11] For example, the genes responsible for testosterone have a beneficial effect early in life, for testosterone promotes reproduction. However, later in life, testosterone accelerates deterioration in arterial walls, suppresses the immune system, and helps cause prostate cancer—changes related to aging.

The antagonistic pleiotropy theory has problems too. First, some genes involved in aging do not have beneficial effects in life. Second, as for those genes that do have beneficial effects early in life and harmful effects later in life, random mutations could produce modifications that eliminate the later harmful effect without losing the early beneficial effect, so one would expect the harmful later effects to disappear from the population as individuals who lacked them outbred their competitors.

The disposable soma theory

The third theory is Thomas Kirkwood's *disposable soma theory*. This theory starts from the premise that an organism has a limited amount of energy to allocate between maintaining its tissues against the basic processes of aging and everything else: courtship, breeding, hunting, food gathering, fighting, and fleeing. Animals whose tissues renew themselves to maintain a perpetually youthful body are using energy that could otherwise be used for reproduction, fighting, and so on. Because natural selection favors species that reproduce more, it favors species whose individuals devote just enough energy to repairing their bodies to live long enough to reproduce as much as those individuals can in their environment—and no more. Tom Kirkwood calls this the "disposable soma" theory because the soma (a "soma" is a body) is a disposable vehicle for combining and distributing the germline cells (eggs and sperm)—somewhat like using up a booster rocket to get a capsule into space. Once that task is accomplished, there is no further need for the soma.[12]

The disposable soma theory faces obstacles as well. First, there is the caloric restriction effect: animals live longer on low-calorie diets with sufficient nutrition. Although nutrition may not be reduced, metabolic energy is reduced, and that should leave less metabolic energy for repair—yet the animals live longer, not shorter lives. Second, females need more metabolic energy for reproduction than males (for gestation and other reasons), yet males in many species live roughly as long as females.

The problems discussed above do not necessarily refute these three theories, and they each have many supporters. They are also compatible; it may turn out that some genes exhibit antagonistic pleiotropy, while others are late-life defects that natural selection does not weed out, and that organisms evolved to conserve metabolic energy for purposes other than needless longevity.

The programmed aging theory

The fourth evolutionary theory of aging is the *programmed aging theory*. This is the earliest of our four theories.[13] In 1889, August Weismann suggested that aging evolved in order to make room for new generations; if there was no room for new generations, then a species would die out. However, this theory has fallen out of favor, for it faces several serious objections. First, animals in captivity live longer and exhibit aging, but in the wild, they don't live long enough to visibly age. If aging is triggered by a genetic program of some kind, we would expect all animals to die when the program kicks in, regardless of where they live. Second, in some animals (including us) there's a late-life mortality plateau. Humans stop aging in their mid-90s. This suggests that the aging program shuts off after a certain age, but that means that those survivors are not being cleared away to make room for new generations. Third, random mutations could produce individuals who either lack the program or have some way of shutting it down and living longer. If they live longer, they will have more offspring and spread their genes through the population, eventually eliminating the death program from the gene pool. However, that seems not to happen.

Recently the programmed aging theory has been revived (though it remains a minority view) using recent developments in evolutionary theory concerning group selection and evolvability.[14] Group selection theory says that evolution sometimes selects for traits that benefit a group but harm individuals, such as altruism. For example, a population that regulates its own population is less susceptible to mass famine or mass epidemic, for it won't overpopulate its habitat as quickly, thus leaving more resources for times of famine or epidemic (this is group selection). Evolvability theory says that evolution sometimes selects for traits that are disadvantageous to an individual but offset that disadvantage by facilitating the individual's ability to evolve. For example, an animal that's older and more experienced but less endowed with intelligence or immunities may survive better than a younger, less experienced animal that has more of those qualities. However, if the older animals in that population die off, then the younger ones acquire experience while retaining their intelligence or immunity. This produces a more fit population overall—thus populations programmed to age may be reproductively more fit. Similarly, if individuals in a population are programmed to age after a certain span of time, the population will evolve faster because there is more room for new generations and hence new mutations. The upshot is that evolution selects for aging, not as a by-product or accident, but because it's advantageous to a population.

The programmed aging theory predicts that there are genetic programs that induce aging. There is some evidence that such programs exist. First, there is apoptosis—programs that kill individual cells that are infected, cancerous, or need to be removed during development (like the webbing between fingers and toes during fetal development). Apoptosis occurs more often late in life, also killing healthy cells. Second, there is replicative senescence—some cells can divide only a limited number of times before their telomeres (genetic caps on the ends of DNA, like the caps on your shoelaces) are too short to allow further division, and then they die because the cell cannot replicate its chromosomes without cutting into its genes. When telomeres get short enough that these cells stop dividing, they seem somehow to alter their gene expression to permit oxidative damage and inhibit cell repair.

These theories have different implications for life extension

These four theories of aging have different implications for the possibility of life extension. The mutational accumulation theory suggests that the mechanisms of aging are quite various and that slowing or halting aging cannot be achieved by tweaking our supposed natural defenses against aging, for it doesn't imply that such defenses exist. It implies that there is no control panel for aging—just a collection of maladies that require a collection of remedies. (By "control panel" I mean a limited set of genetic mechanisms that control all aspects of aging.) That doesn't make life extension impossible, just more difficult to achieve. The antagonistic pleiotropy theory faces that problem too, for it says that aging results from a variety of genes with beneficial effects early in life and harmful effects later on.

The prospects for life extension look brighter on the disposable soma theory and the programmed aging theory. According to the disposable soma theory, aging results from a failure of the maintenance systems we already have. It's possible to slow or halt aging because the body already does, at least during youth, and to a lesser extent later on. Moreover, aging occurs all over the body at roughly the same time in multiple organs and tissues, which suggests one set of maintenance systems. This is also true of the programmed aging theory, except that instead of systems that slow or halt aging, we have systems that cause it—again, all over the body at roughly the same time, which suggests one set of aging systems. On both of these theories, there may be fewer basic processes of aging and thus a control panel for aging.

Of course, it may also turn out that all four theories contain part of the truth and that there is a control panel for some aspects of aging but not for others, thereby complicating the search for life extension.

Appendix B

Bernard Williams, Personal Identity, and Categorical Desires

B.1 Williams's third life extension scenario

When Bernard Williams discusses whether immortality is desirable, he considers three possible scenarios. In the first, you don't change, and you eventually become intolerably bored. In the second, you change enough to acquire new interests and avoid boredom, but at the price of eventually changing so much that you evolve into a new person, and thereby cease to exist. In the third scenario, you change just as much as you do in the second, but you don't cease to exist (not because you change in a different way, but for other reasons). Williams thinks immortality is undesirable in the first scenario, impossible in the second, and undesirable in the third.

In chapter 2, I discussed only the first two scenarios. I did this because the objection I consider successful against the second scenario also scores against the third and because the third scenario is quite involved and blows us somewhat off course. I will discuss it here.

As I said, in the third scenario you change enough to avoid permanent and intolerable boredom, and yet you don't cease to exist. How is it that you change so much but still exist? The reason is not that there's something different about you. The reason is that the right theory of personal identity is different than most philosophers think it is. Most philosophers who have thought about personal identity over time have subscribed to some version of the theory that you still exist only if your later self has enough continuity and connectedness with your earlier self. To put it simply, you must have enough psychological features in common with your earlier self. Williams seems to think that this theory is correct, but he also considers the possibility that he and others are wrong and that some other theory of personal identity is correct.

On other theories, you can continue to exist even while changing dramatically over time. This might happen if, for example, you have an immortal soul that is not the same as your personality and temperament, and your survival is simply a matter of that soul's survival, or if (according to another theory) you still exist as long as your brain does, regardless of your psychological properties over time. If one of those theories is true, then you can be immortal and avoid permanent and intolerable boredom. Williams wants to show that immortality is not desirable even if one of these theories is correct, so he asks whether that kind of existence is desirable. This is not the same question as whether you continue to exist; it's a question about whether you would want to continue existing in that way: "That is, in my view, a different question from the question of whether it will be him" even after centuries of change.[1]

B.2 Categorical desires and why we want to keep on living

Williams thinks you have no reason to desire a future in which you continue to exist forever but evade intolerable boredom only by changing your character, personality, and values dramatically over time. He doesn't claim it would be an unpleasant life, just not one that it makes any sense to desire. Why not? Williams's answer to this question starts with his answer to another question: When, in general, do we have reason to continue living? It's not enough that we desire simply to keep on living—"that desire itself . . . has to be sustained or filled out by some desire for something else, even if it is only, at the margin, the desire that future desires of mine will be born and satisfied."[2] In other words, it makes no sense to say that you simply desire to be alive. You desire to continue living because you need to remain alive in order to satisfy some other desires you have or will have.[3] We must desire that our continued life have features that make it worth living, and we must desire those features.

According to Williams, desiring those features involves a special kind of desire. He distinguishes between desires that are conditioned on your being alive and desires that are not conditioned on your being alive, which he calls "categorical desires." A desire for painkillers would be a conditional desire: getting such painkillers is not a reason to desire to continue living, but if you're going to be here anyway, you want to be free of pain. Presumably a categorical desire would be something like writing a book you've always wanted to write—it gives you a reason to want to go on at least long enough to see it in print. According to Williams, the more categorical desires you have (and perhaps the stronger they are), the more reason you have to keep on living.[4]

A categorical desire is a reason for me to keep on living. A desire can't play that role unless the object of the desire—whatever fulfills the desire, in other words—requires that I be there when it happens. That can happen either because (a) my participation will help bring about the object of that desire (I want to mentor my niece so that she graduates from high school on time), or because (b) my desire for that object includes the desire that I be there to see it happen (I want to watch her graduate), or both.

Williams doesn't mention (a) and (b), but I think they represent what he has in mind, for they are crucial to why I have no reason to desire Williams's third scenario of what immortality would be like—that is, no reason to desire to remain alive forever while endlessly evolving my psyche so much that I seem like a different person, even though (unlike the second scenario) I still exist. There are two conditions for me to rationally desire that I continue existing. First, whoever is living that future life must clearly be me. That makes sense; if the survivor isn't me, then I didn't get the future I desired. On the theories of personal identity assumed under the third scenario, this condition is met.

The second condition—this is where (a) and (b) come in—is that "the state in which I survive should be one that, to me looking forward, will be adequately related, in the life it presents, to those aims I now have in wanting to survive at all." He also states the second condition this way: "What is promised must hold out some hopes for" my current categorical desires.[5] This takes us back to (a) and (b): I need to be there, in that future, to help bring about the fulfillment of my categorical desires (they wouldn't be categorical if I didn't). That will happen only if my future self has the same categorical desires I have. If he lacks my categorical desire for X, then his existence does nothing to improve the odds that X will happen, and I therefore have no reason to desire that he (I) will exist. In other words, for my future state to be "adequately related" to my present categorical desires (Williams's second condition), I must have those desires in that future. It is therefore not enough that my future self is still me; he must share my categorical desires. However, to avoid intolerable boredom, he must also be so different from me that he has different categorical desires. That means he will not share my current categorical desires, either because he can't remember the earlier phase of life when he had them or because, even though he remembers having them, he no longer cares.[6] Because his existence doesn't promote my current categorical desires, I have no reason to desire that he exists, even though he is me.[7] Therefore, I have no reason to desire the third scenario, even though I would still exist.

B.3 The Tarzan objection to the third scenario

As I said, Williams's argument about the third scenario, like his argument about the second, falls victim to the Tarzan objection. With respect to the second scenario, the Tarzan objection was that I have reason to care about myself in the next 20 years, and 20 years from now, I will have reason to care about my future self 20 years beyond that (imagine Tarzan taking 20 years to swing from one vine to the next), so even if I evolve into a new person and cease to exist, at any given time, I have reason to want to live a while longer. The Tarzan objection works almost the same way for the third scenario.

In the third scenario, the puzzle is not why I might care about my future self when he is not me. Rather, the puzzle is why I might care about my future self even though he doesn't share my categorical desires. The answer is much the same: I may not have reason to care about my future self 200 years from now, for he won't care about my desires for that future, but I do have reason to care about my future self 20 years from now, for he will share some (but not all) of my categorical desires to some extent. Stage by stage, like swinging from vine to vine, I have reason to keep going.

B.4 Two unsuccessful objections to the third scenario

In the rest of this appendix, I want to consider some objections to Williams's arguments about the third scenario that, in my judgment, are not successful. Although I think his conclusion is mistaken (mainly because of the Tarzan objection), I also believe that his arguments are stronger and more subtle than they might appear to be.

The first is an objection that I could have a categorical desire that my future self's categorical desires be fulfilled. In other words, I don't share the categorical desires that I'll have 500 years from now, but nonetheless,

I want those desires fulfilled just because they will be mine. Williams concedes that this is possible as a "limiting case."[8]

Of course, that limiting case sounds a lot like a desire to keep on living as long as you enjoy being alive, even knowing that you will change radically over time. If that's possible, then you could have reason to desire a kind of immortality. However, Williams does not pursue this issue and seems not to be worried about it. Why not? Perhaps for this reason: he claims that my only reason for wanting to continue to exist is that my future existence promotes the fulfillment of desires I now have. If that is right, then I can't desire that I flourish in the future unless that fulfills one of my current categorical desires for something other than my existence. My future categorical desires could be (and eventually will be, if I live long enough) wildly different from anything I desire now. I therefore have no reason to desire their fulfillment based on the content of those desires. What reason do I have, then? The mere fact that they will one day be mine? That makes sense only if I have a reason to care about my future self, regardless of whether he shares my current categorical desires. However, according to Williams, I can't have reason to care about my future self unless his existence promotes my current categorical desires; therefore, I don't have a reason to care about my future self regardless of whether he shares my categorical desires. We seem to be caught in a circle, so the objection fails.

The second unsuccessful objection is that I can desire that my distant, radically different future self continue to exist not because I have any desires about his desires but simply because I believe that his existence could be good for him whether he wants it or not. Larry Temkin argues that immortality could be good for you even independently of whether you desire it if living forever is best for you in some objective way that is not tied to your desires.[9] For example, life might be good for my future self (who is still me, by the way), objectively and independent of what he wants, if it contains certain things to a substantial degree: moral goodness, rational activity, developing his abilities, having children and being a good parent, knowledge, and awareness of beauty, among other things (along with the absence of various bad things).[10] Let's assume this theory is correct and that my future selves are better off if their lives contain these things.

Certainly you could have a categorical desire that your future self be objectively better off in some way that bears no relation to his categorical desires. But why would you? Because you care about your future self simply by virtue of his being you? Williams is right: that's not enough. We need categorical desires to make sense of the desire to keep on living. You could have a categorical desire that your future self will be objectively well off, but that's not a self-interested desire that you continue to exist unless the fact that he is you plays a role in that desire. In William's account, you care about him because he shares your categorical desires and will therefore help fulfill them. In Temkin's account, you have no such reason to care about him, so Temkin needs to provide some other reason why you care so much about it. That reason cannot simply be that he is you, for that begs the question. We are looking for a reason why you should care that he still exists—that is, why you should want to live that long. Williams provides one, but Temkin does not.

Appendix C

Demographic Tables and Graphs

These are the tables and graphs referred to in chapter 6. They were prepared by Shahin Davoudpour, a doctoral candidate in the school of social sciences at the University of California, Irvine. There are 13 tables and 13 graphs illustrating those tables.

Table 1.

150-year life expectancy, 1.0 child at average age 25, 1.0 child at average age 75

Generation	Year	Birth at 25	Birth at 75	Population
0	0	0	0	1000
1	25	500	0	1500
2	50	250	0	1750
3	75	125	500	2375
4	100	312.5	250	2937.5
5	125	281.25	125	3343.75
6	150	203.125	312.5	2859.375
7	175	257.8125	281.25	2898.4375
8	200	269.5313	203.125	3121.0938
9	225	236.3281	257.8125	2990.2344
10	250	247.0703	269.53125	2944.3359
11	275	258.3008	236.328125	3032.7148
12	300	247.3145	247.0703125	3011.4746
13	325	247.1924	258.3007813	2977.9053
14	350	252.7466	247.3144531	3005.3101
15	375	250.0305	247.1923828	3008.3923
16	400	248.6115	252.746582	2993.1488
17	425	250.679	250.0305176	2999.2294
18	450	250.3548	248.6114502	3003.8109
19	475	249.4831	250.6790161	2998.4798
20	500	250.0811	250.3547668	2998.8546
21	525	250.2179	249.4831085	3001.3328
22	550	249.8505	250.0810623	2999.9063
23	575	249.9658	250.2179146	2999.3805
24	600	250.0919	249.8505116	3000.3566
25	625	249.9712	249.9657869	3000.1315
26	650	249.9685	250.0918508	2999.756
27	675	250.0302	249.9711812	3000.0563
28	700	250.0007	249.968484	3000.0939
29	725	249.9846	250.0301674	2999.9249
30	750	250.0074	250.0006743	2999.9906
31	775	250.004	249.9845792	3000.0422
32	800	249.9943	250.0073733	2999.9836

(continued)

Table 1. (*continued*)

Generation	Year	Birth at 25	Birth at 75	Population
33	825	250.0008	250.0040238	2999.9871
34	850	250.0024	249.9943015	3000.0147
35	875	249.9984	250.0008374	2999.9991
36	900	249.9996	250.0024306	2999.9931
37	925	250.001	249.998366	3000.0039
38	950	249.9997	249.9996017	3000.0015
39	975	249.9996	250.0010161	2999.9973
40	1000	250.0003	249.9996911	3000.0006
41	1025	250	249.9996464	3000.001
42	1050	249.9998	250.0003313	2999.9992
43	1075	250.0001	250.0000112	2999.9999
44	1100	250	249.9998288	3000.0005
45	1125	249.9999	250.00008	2999.9998
46	1150	250	250.0000456	2999.9999
47	1175	250	249.9999372	3000.0002
48	1200	250	250.0000086	3000
49	1225	250	250.0000271	2999.9999
50	1250	250	249.9999821	3000
51	1275	250	249.9999954	3000
52	1300	250	250.0000112	3000
53	1325	250	249.9999967	3000
54	1350	250	249.999996	3000
55	1375	250	250.0000036	3000
56	1400	250	250.0000002	3000
57	1425	250	249.9999981	3000
58	1450	250	250.0000009	3000
59	1475	250	250.0000005	3000
60	1500	250	249.9999993	3000
61	1525	250	250.0000001	3000
62	1550	250	250.0000003	3000
63	1575	250	249.9999998	3000
64	1600	250	249.9999999	3000
65	1625	250	250.0000001	3000
66	1650	250	250	3000

(*continued*)

Table 1. (*continued*)

Generation	Year	Birth at 25	Birth at 75	Population
67	1675	250	250	3000
68	1700	250	250	3000
69	1725	250	250	3000
70	1750	250	250	3000
71	1775	250	250	3000
72	1800	250	250	3000
73	1825	250	250	3000
74	1850	250	250	3000
75	1875	250	250	3000
76	1900	250	250	3000
77	1925	250	250	3000
78	1950	250	250	3000
79	1975	250	250	3000
80	2000	250	250	3000

Graph 1. Life extension of 150 years, 1.0 child at age 25 and 1.0 child at age 75

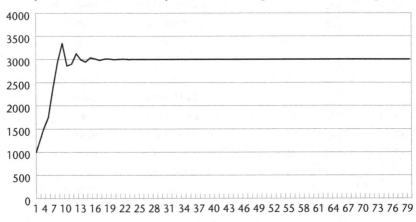

———— Population

Table 2.
150-year life expectancy, 1.0 child at average age 25, 1.0 child at average age 50, 1.0 child at average age 75

Generation	Year	Birth at 25	Birth at 50	Birth at 75	Population
0	0	1000	0	0	1000
1	25	500	0	0	1500
2	50	250	0	0	1750
3	75	125	500	0	2375
4	100	62.5	250	0	2687.5
5	125	156.25	125	0	2968.75
6	150	140.625	312.5	156.25	2578.125
7	175	304.6875	156.25	140.625	2679.6875
8	200	300.7813	140.625	226.5625	3097.6563
9	225	333.9844	304.6875	230.46875	3841.7969
10	250	434.5703	300.78125	220.703125	4735.3516
11	275	478.0273	333.984375	319.3359375	5710.4492
12	300	565.6738	434.5703125	367.6757813	6468.9941
13	325	683.96	478.0273438	406.0058594	7435.4248
14	350	783.9966	565.6738281	500.1220703	8617.2485
15	375	924.8962	683.9599609	580.9936523	9937.9578
16	400	1094.925	783.996582	674.8352051	11535.66
17	425	1276.878	924.8962402	804.4281006	13410.515
18	450	1503.101	1094.924927	939.4607544	15580.082
19	475	1768.744	1276.878357	1100.887299	18158.598
20	500	2073.255	1503.101349	1299.013138	21184.175
21	525	2437.685	1768.743515	1522.810936	24723.564
22	550	2864.619	2073.254585	1788.177967	28895.859
23	575	3363.026	2437.684536	2103.214025	33793.581
24	600	3951.962	2864.619493	2468.937039	39541.613
25	625	4642.759	3363.026023	2900.355279	46301.244
26	650	5453.07	3951.962292	3408.290893	54239.199
27	675	6406.662	4642.759413	4002.892718	63562.274
28	700	7526.157	5453.070357	4702.516325	74517.965
29	725	8840.872	6406.661771	5524.710592	87386.285
30	750	10386.12	7526.156951	6489.613654	102502.66
31	775	12200.95	8840.871817	7623.766794	120262.1

(*continued*)

Table 2. (*continued*)

Generation	Year	Birth at 25	Birth at 50	Birth at 75	Population
32	800	14332.79	10386.12209	8956.13952	141123.83
33	825	16837.53	12200.94635	10520.90908	165630.9
34	850	19779.69	14332.79248	12359.45728	194421.1
35	875	23235.97	16837.52704	14519.2367	228241.59
36	900	27296.37	19779.69124	17056.24186	267972
37	925	32066.15	23235.9705	20036.74877	314645.28
38	950	37669.43	27296.36712	23538.02918	369474.06
39	975	44251.92	32066.15011	27651.0603	433883.8
40	1000	51984.56	37669.43469	32482.90091	509548.76
41	1025	61068.45	44251.91549	38159.0328	598435.42
42	1050	71739.7	51984.56295	44826.99882	702854.38
43	1075	84275.63	61068.44927	52660.18238	825519.78
44	1100	99002.13	71739.69878	61862.13087	969619.91
45	1125	116302	84275.63028	72672.03978	1138900.4
46	1150	136624.8	99002.13097	85370.91488	1337761.4
47	1175	160498.9	116301.9803	100288.8053	1571371.7
48	1200	188544.9	136624.8252	117813.4781	1845803.6
49	1225	221491.6	160498.9355	138400.4579	2168190.3
50	1250	260195.5	188544.8606	162584.8429	2546911.6
51	1275	305662.6	221491.5819	190995.2587	2991811.4
52	1300	359074.7	260195.4877	224370.1741	3514453.9
53	1325	421820.2	305662.5955	263577.0887	4128424
54	1350	495529.9	359074.7181	309635.1029	4849680.6
55	1375	582119.9	421820.1899	363741.3927	5696971.1
56	1400	683840.7	495529.9371	427302.3276	6692318.9
57	1425	803336.5	582119.879	501970.0345	7861595.9
58	1450	943713.2	683840.7309	589685.334	9235194.8
59	1475	1108620	803336.4978	692728.1884	10848819
60	1500	1302342	943713.2056	813776.9683	12744412
61	1525	1529916	1108619.635	955978.0665	14971244
62	1550	1797257	1302342.161	1123027.683	17587198
63	1575	2111313	1529916.167	1319267.901	20660269
64	1600	2480249	1797256.935	1549799.548	24270335
65	1625	2913653	2111313.389	1820614.778	28511231

(continued)

Table 2. (*continued*)

Generation	Year	Birth at 25	Birth at 50	Birth at 75	Population
66	1650	3422790	2480248.729	2138752.832	33493191
67	1675	4020896	2913652.606	2512482.997	39345709
68	1700	4723516	3422790.386	2951519.558	46220908
69	1725	5548913	4020895.974	3467274.29	54297493
70	1750	6518542	4723515.788	4073153.087	63785399
71	1775	7657605	5548912.866	4784904.42	74931240
72	1800	8995711	6518541.565	5621028.676	88024730
73	1825	10567641	7657605.22	6603259.043	103406203
74	1850	12414253	8995711.253	7757126.409	121475468
75	1875	14583545	10567640.75	9112622.984	142702193
76	1900	17131904	12414252.51	10704981.88	167638122
77	1925	20125569	14583545.08	12575592.92	196931407
78	1950	23642354	17131904.41	14773078.46	231343462
79	1975	27773668	20125569.4	17354557.24	271768752
80	2000	32626897	23642353.7	20387129.05	319258042

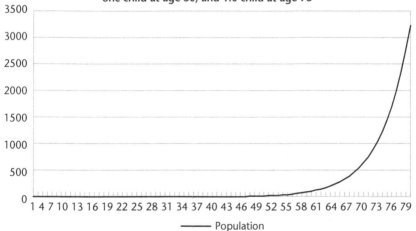

Graph 2. Life extension of 150 years, 1.0 child at age 25,
one child at age 50, and 1.0 child at age 75

——— Population

Table 3.
150-year life expectancy, 1.0 child at average age 25, 0.5 child at average age 75

Generation	Year	Birth at 25	Birth at 75	Population
0	0	0	0	1000
1	25	500	0	1500
2	50	250	0	1750
3	75	125	250	2125
4	100	187.5	125	2437.5
5	125	156.25	62.5	2656.25
6	150	109.375	93.75	1859.375
7	175	101.5625	78.125	1539.0625
8	200	89.84375	54.6875	1433.5938
9	225	72.26563	50.78125	1181.6406
10	250	61.52344	44.921875	975.58594
11	275	53.22266	36.1328125	846.19141
12	300	44.67773	30.76171875	718.50586
13	325	37.71973	26.61132813	603.14941
14	350	32.16553	22.33886719	513.12256
15	375	27.2522	18.85986328	436.18774
16	400	23.05603	16.08276367	368.88123
17	425	19.5694	13.62609863	312.72125
18	450	16.59775	11.52801514	265.40756
19	475	14.06288	9.784698486	224.92409
20	500	11.92379	8.298873901	190.64236
21	525	10.11133	7.031440735	161.67307
22	550	8.571386	5.961894989	137.06756
23	575	7.266641	5.05566597	116.19437
24	600	6.161153	4.285693169	98.515451
25	625	5.223423	3.633320332	83.524615
26	650	4.428372	3.080576658	70.810899
27	675	3.754474	2.611711621	60.034312
28	700	3.183093	2.214185894	50.89831
29	725	2.698639	1.877237111	43.15188
30	750	2.287938	1.591546461	36.584518
31	775	1.939742	1.349319704	31.016836
32	800	1.644531	1.14396913	26.296388

(continued)

Table 3. (*continued*)

Generation	Year	Birth at 25	Birth at 75	Population
33	825	1.39425	0.96987118	22.294324
34	850	1.182061	0.822265516	18.901371
35	875	1.002163	0.69712504	16.024782
36	900	0.849644	0.591030315	13.585972
37	925	0.720337	0.501081537	11.518329
38	950	0.610709	0.424822028	9.76536
39	975	0.517766	0.360168593	8.279173
40	1000	0.438967	0.305354681	7.0191687
41	1025	0.372161	0.258882847	5.9509244
42	1050	0.315522	0.219483572	5.0452554
43	1075	0.267503	0.186080456	4.2774199
44	1100	0.226792	0.15776094	3.626441
45	1125	0.192276	0.133751363	3.0745344
46	1150	0.163014	0.113395796	2.6066222
47	1175	0.138205	0.096138133	2.2099213
48	1200	0.117171	0.081506907	1.8735943
49	1225	0.099339	0.069102402	1.5884527
50	1250	0.084221	0.058585734	1.3467067
51	1275	0.071403	0.049669594	1.1417519
52	1300	0.060536	0.042110398	0.9679891
53	1325	0.051323	0.035701632	0.8206712
54	1350	0.043513	0.030268215	0.6957736
55	1375	0.03689	0.025661707	0.5898841
56	1400	0.031276	0.021756261	0.5001098
57	1425	0.026516	0.018445184	0.4239983
58	1450	0.022481	0.015638019	0.3594702
59	1475	0.019059	0.013258075	0.3047626
60	1500	0.016159	0.011240334	0.2583809
61	1525	0.0137	0.009529671	0.219058
62	1550	0.011615	0.008079354	0.1857196
63	1575	0.009847	0.006849761	0.157455
64	1600	0.008348	0.005807298	0.133492
65	1625	0.007078	0.004923488	0.1131759
66	1650	0.006001	0.004174184	0.0959517

(*continued*)

Table 3. (*continued*)

Generation	Year	Birth at 25	Birth at 75	Population
67	1675	0.005087	0.003538917	0.0813489
68	1700	0.004313	0.00300033	0.0689684
69	1725	0.003657	0.002543711	0.0584721
70	1750	0.0031	0.002156585	0.0495733
71	1775	0.002628	0.001828375	0.0420287
72	1800	0.002228	0.001550115	0.0356324
73	1825	0.001889	0.001314204	0.0302095
74	1850	0.001602	0.001114196	0.0256119
75	1875	0.001358	0.000944627	0.0217141
76	1900	0.001151	0.000800864	0.0184094
77	1925	0.000976	0.000678981	0.0156077
78	1950	0.000828	0.000575647	0.0132324
79	1975	0.000702	0.00048804	0.0112185
80	2000	0.000595	0.000413765	0.0095112

Graph 3. Life extension of 150 years, 1.0 child at age 25 and 0.5 child at age 500

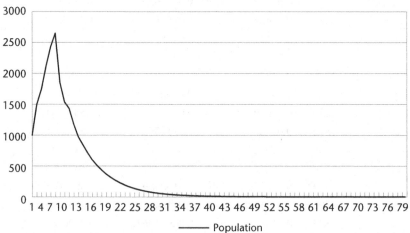

Population

Table 4.
150-year life expectancy, 1.0 child at average age 25

Generation	Year	Birth at 25	Population
0	0	0	1000
1	25	500	1500
2	50	250	1750
3	75	125	1875
4	100	62.5	1937.5
5	125	31.25	1968.75
6	150	15.625	984.375
7	175	7.8125	492.1875
8	200	3.90625	246.09375
9	225	1.953125	123.04688
10	250	0.976563	61.523438
11	275	0.488281	30.761719
12	300	0.244141	15.380859
13	325	0.12207	7.6904297
14	350	0.061035	3.8452148
15	375	0.030518	1.9226074
16	400	0.015259	0.9613037
17	425	0.007629	0.4806519
18	450	0.003815	0.2403259
19	475	0.001907	0.120163
20	500	0.000954	0.0600815
21	525	0.000477	0.0300407
22	550	0.000238	0.0150204
23	575	0.000119	0.0075102
24	600	5.96E-05	0.0037551
25	625	2.98E-05	0.0018775
26	650	1.49E-05	0.0009388
27	675	7.45E-06	0.0004694
28	700	3.73E-06	0.0002347
29	725	1.86E-06	0.0001173
30	750	9.31E-07	5.867E-05
31	775	4.66E-07	2.934E-05
32	800	2.33E-07	1.467E-05

(*continued*)

Table 4. (*continued*)

Generation	Year	Birth at 25	Population
33	825	1.16E-07	7.334E-06
34	850	5.82E-08	3.667E-06
35	875	2.91E-08	1.834E-06
36	900	1.46E-08	9.168E-07
37	925	7.28E-09	4.584E-07
38	950	3.64E-09	2.292E-07
39	975	1.82E-09	1.146E-07
40	1000	9.09E-10	5.73E-08
41	1025	4.55E-10	2.865E-08
42	1050	2.27E-10	1.432E-08
43	1075	1.14E-10	7.162E-09
44	1100	5.68E-11	3.581E-09
45	1125	2.84E-11	1.791E-09
46	1150	1.42E-11	8.953E-10
47	1175	7.11E-12	4.476E-10
48	1200	3.55E-12	2.238E-10
49	1225	1.78E-12	1.119E-10
50	1250	8.88E-13	5.596E-11
51	1275	4.44E-13	2.798E-11
52	1300	2.22E-13	1.399E-11
53	1325	1.11E-13	6.994E-12
54	1350	5.55E-14	3.497E-12
55	1375	2.78E-14	1.749E-12
56	1400	1.39E-14	8.743E-13
57	1425	6.94E-15	4.372E-13
58	1450	3.47E-15	2.186E-13
59	1475	1.73E-15	1.093E-13
60	1500	8.67E-16	5.464E-14
61	1525	4.34E-16	2.732E-14
62	1550	2.17E-16	1.366E-14
63	1575	1.08E-16	6.83E-15
64	1600	5.42E-17	3.415E-15
65	1625	2.71E-17	1.708E-15
66	1650	1.36E-17	8.538E-16

(*continued*)

Table 4. (*continued*)

Generation	Year	Birth at 25	Population
67	1675	6.78E-18	4.269E-16
68	1700	3.39E-18	2.135E-16
69	1725	1.69E-18	1.067E-16
70	1750	8.47E-19	5.336E-17
71	1775	4.24E-19	2.668E-17
72	1800	2.12E-19	1.334E-17
73	1825	1.06E-19	6.67E-18
74	1850	5.29E-20	3.335E-18
75	1875	2.65E-20	1.668E-18
76	1900	1.32E-20	8.338E-19
77	1925	6.62E-21	4.169E-19
78	1950	3.31E-21	2.084E-19
79	1975	1.65E-21	1.042E-19
80	2000	8.27E-22	5.211E-20

Graph 4. Life extension of 150 years, 1.0 child at age 25

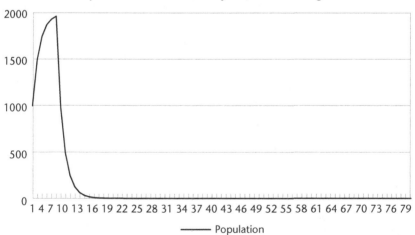

— Population

Table 5.
150-year life expectancy, 1.0 child at average age 50

Generation	Year	Birth at 50	Population
0	0	0	1000
1	50	500	1500
2	100	250	1750
3	150	125	875
4	200	62.5	437.5
5	250	31.25	218.75
6	300	15.625	109.375
7	350	7.8125	54.6875
8	400	3.90625	27.34375
9	450	1.953125	13.671875
10	500	0.9765625	6.8359375
11	550	0.48828125	3.4179688
12	600	0.244140625	1.7089844
13	650	0.122070313	0.8544922
14	700	0.061035156	0.4272461
15	750	0.030517578	0.213623
16	800	0.015258789	0.1068115
17	850	0.007629395	0.0534058
18	900	0.003814697	0.0267029
19	950	0.001907349	0.0133514
20	1000	0.000953674	0.0066757
21	1050	0.000476837	0.0033379
22	1100	0.000238419	0.0016689
23	1150	0.000119209	0.0008345
24	1200	5.96046E-05	0.0004172
25	1250	2.98023E-05	0.0002086
26	1300	1.49012E-05	0.0001043
27	1350	7.45058E-06	5.215E-05
28	1400	3.72529E-06	2.608E-05
29	1450	1.86265E-06	1.304E-05
30	1500	9.31323E-07	6.519E-06
31	1550	4.65661E-07	3.26E-06
32	1600	2.32831E-07	1.63E-06

(*continued*)

Table 5. (*continued*)

Generation	Year	Birth at 50	Population
33	1650	1.16415E-07	8.149E-07
34	1700	5.82077E-08	4.075E-07
35	1750	2.91038E-08	2.037E-07
36	1800	1.45519E-08	1.019E-07
37	1850	7.27596E-09	5.093E-08
38	1900	3.63798E-09	2.547E-08
39	1950	1.81899E-09	1.273E-08
40	2000	9.09495E-10	6.366E-09
41	2050	4.54747E-10	3.183E-09
42	2100	2.27374E-10	1.592E-09
43	2150	1.13687E-10	7.958E-10
44	2200	5.68434E-11	3.979E-10
45	2250	2.84217E-11	1.99E-10
46	2300	1.42109E-11	9.948E-11
47	2350	7.10543E-12	4.974E-11
48	2400	3.55271E-12	2.487E-11
49	2450	1.77636E-12	1.243E-11
50	2500	8.88178E-13	6.217E-12
51	2550	4.44089E-13	3.109E-12
52	2600	2.22045E-13	1.554E-12
53	2650	1.11022E-13	7.772E-13
54	2700	5.55112E-14	3.886E-13
55	2750	2.77556E-14	1.943E-13
56	2800	1.38778E-14	9.714E-14
57	2850	6.93889E-15	4.857E-14
58	2900	3.46945E-15	2.429E-14
59	2950	1.73472E-15	1.214E-14
60	3000	8.67362E-16	6.072E-15
61	3050	4.33681E-16	3.036E-15
62	3100	2.1684E-16	1.518E-15
63	3150	1.0842E-16	7.589E-16
64	3200	5.42101E-17	3.795E-16
65	3250	2.71051E-17	1.897E-16
66	3300	1.35525E-17	9.487E-17

(*continued*)

Table 5. (*continued*)

Generation	Year	Birth at 50	Population
67	3350	6.77626E-18	4.743E-17
68	3400	3.38813E-18	2.372E-17
69	3450	1.69407E-18	1.186E-17
70	3500	8.47033E-19	5.929E-18
71	3550	4.23516E-19	2.965E-18
72	3600	2.11758E-19	1.482E-18
73	3650	1.05879E-19	7.412E-19
74	3700	5.29396E-20	3.706E-19
75	3750	2.64698E-20	1.853E-19
76	3800	1.32349E-20	9.264E-20
77	3850	6.61744E-21	4.632E-20
78	3900	3.30872E-21	2.316E-20
79	3950	1.65436E-21	1.158E-20
80	4000	8.27181E-22	5.79E-21

Graph 5. Life extension of 150 years, 1.0 child at age 50

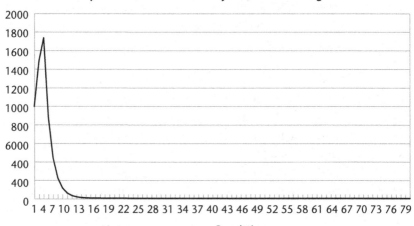

Table 6.
150-year life expectancy, 0.5 child at average age 25

Generation	Year	Birth at 25	Population
0	0	0	1000
1	25	250	1250
2	50	62.5	1312.5
3	75	15.625	1328.125
4	100	3.90625	1332.0313
5	125	0.976563	1333.0078
6	150	0.244141	333.25195
7	175	0.061035	83.312988
8	200	0.015259	20.828247
9	225	0.003815	5.2070618
10	250	0.000954	1.3017654
11	275	0.000238	0.3254414
12	300	5.96E-05	0.0813603
13	325	1.49E-05	0.0203401
14	350	3.73E-06	0.005085
15	375	9.31E-07	0.0012713
16	400	2.33E-07	0.0003178
17	425	5.82E-08	7.945E-05
18	450	1.46E-08	1.986E-05
19	475	3.64E-09	4.966E-06
20	500	9.09E-10	1.241E-06
21	525	2.27E-10	3.104E-07
22	550	5.68E-11	7.759E-08
23	575	1.42E-11	1.94E-08
24	600	3.55E-12	4.849E-09
25	625	8.88E-13	1.212E-09
26	650	2.22E-13	3.031E-10
27	675	5.55E-14	7.577E-11
28	700	1.39E-14	1.894E-11
29	725	3.47E-15	4.736E-12
30	750	8.67E-16	1.184E-12
31	775	2.17E-16	2.96E-13
32	800	5.42E-17	7.4E-14

(continued)

Table 6. (*continued*)

Generation	Year	Birth at 25	Population
33	825	1.36E-17	1.85E-14
34	850	3.39E-18	4.625E-15
35	875	8.47E-19	1.156E-15
36	900	2.12E-19	2.89E-16
37	925	5.29E-20	7.226E-17
38	950	1.32E-20	1.807E-17
39	975	3.31E-21	4.516E-18
40	1000	8.27E-22	1.129E-18
41	1025	2.07E-22	2.823E-19
42	1050	5.17E-23	7.057E-20
43	1075	1.29E-23	1.764E-20
44	1100	3.23E-24	4.411E-21
45	1125	8.08E-25	1.103E-21
46	1150	2.02E-25	2.757E-22
47	1175	5.05E-26	6.891E-23
48	1200	1.26E-26	1.723E-23
49	1225	3.16E-27	4.307E-24
50	1250	7.89E-28	1.077E-24
51	1275	1.97E-28	2.692E-25
52	1300	4.93E-29	6.73E-26
53	1325	1.23E-29	1.682E-26
54	1350	3.08E-30	4.206E-27
55	1375	7.7E-31	1.052E-27
56	1400	1.93E-31	2.629E-28
57	1425	4.81E-32	6.572E-29
58	1450	1.2E-32	1.643E-29
59	1475	3.01E-33	4.108E-30
60	1500	7.52E-34	1.027E-30
61	1525	1.88E-34	2.567E-31
62	1550	4.7E-35	6.418E-32
63	1575	1.18E-35	1.605E-32
64	1600	2.94E-36	4.011E-33
65	1625	7.35E-37	1.003E-33
66	1650	1.84E-37	2.507E-34

(*continued*)

Table 6. (*continued*)

Generation	Year	Birth at 25	Population
67	1675	4.59E-38	6.268E-35
68	1700	1.15E-38	1.567E-35
69	1725	2.87E-39	3.917E-36
70	1750	7.17E-40	9.793E-37
71	1775	1.79E-40	2.448E-37
72	1800	4.48E-41	6.121E-38
73	1825	1.12E-41	1.53E-38
74	1850	2.8E-42	3.826E-39
75	1875	7.01E-43	9.564E-40
76	1900	1.75E-43	2.391E-40
77	1925	4.38E-44	5.977E-41
78	1950	1.09E-44	1.494E-41
79	1975	2.74E-45	3.736E-42
80	2000	6.84E-46	9.34E-43

Graph 6. Life extension of 150 years, 0.5 child at age 25

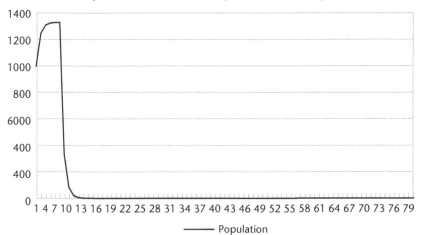

— Population

Table 7.
150-year life expectancy, 0.5 child at average age 50

Generation	Year	Birth at 50	Population
0	0	0	1000
1	50	250	1250
2	100	62.5	1312.5
3	150	15.625	328.125
4	200	3.90625	82.03125
5	250	0.9765625	20.507813
6	300	0.244140625	5.1269531
7	350	0.061035156	1.2817383
8	400	0.015258789	0.3204346
9	450	0.003814697	0.0801086
10	500	0.000953674	0.0200272
11	550	0.000238419	0.0050068
12	600	5.96046E-05	0.0012517
13	650	1.49012E-05	0.0003129
14	700	3.72529E-06	7.823E-05
15	750	9.31323E-07	1.956E-05
16	800	2.32831E-07	4.889E-06
17	850	5.82077E-08	1.222E-06
18	900	1.45519E-08	3.056E-07
19	950	3.63798E-09	7.64E-08
20	1000	9.09495E-10	1.91E-08
21	1050	2.27374E-10	4.775E-09
22	1100	5.68434E-11	1.194E-09
23	1150	1.42109E-11	2.984E-10
24	1200	3.55271E-12	7.461E-11
25	1250	8.88178E-13	1.865E-11
26	1300	2.22045E-13	4.663E-12
27	1350	5.55112E-14	1.166E-12
28	1400	1.38778E-14	2.914E-13
29	1450	3.46945E-15	7.286E-14
30	1500	8.67362E-16	1.821E-14
31	1550	2.1684E-16	4.554E-15
32	1600	5.42101E-17	1.138E-15

(continued)

Table 7. (*continued*)

Generation	Year	Birth at 50	Population
33	1650	1.35525E-17	2.846E-16
34	1700	3.38813E-18	7.115E-17
35	1750	8.47033E-19	1.779E-17
36	1800	2.11758E-19	4.447E-18
37	1850	5.29396E-20	1.112E-18
38	1900	1.32349E-20	2.779E-19
39	1950	3.30872E-21	6.948E-20
40	2000	8.27181E-22	1.737E-20
41	2050	2.06795E-22	4.343E-21
42	2100	5.16988E-23	1.086E-21
43	2150	1.29247E-23	2.714E-22
44	2200	3.23117E-24	6.785E-23
45	2250	8.07794E-25	1.696E-23
46	2300	2.01948E-25	4.241E-24
47	2350	5.04871E-26	1.06E-24
48	2400	1.26218E-26	2.651E-25
49	2450	3.15544E-27	6.626E-26
50	2500	7.88861E-28	1.657E-26
51	2550	1.97215E-28	4.142E-27
52	2600	4.93038E-29	1.035E-27
53	2650	1.2326E-29	2.588E-28
54	2700	3.08149E-30	6.471E-29
55	2750	7.70372E-31	1.618E-29
56	2800	1.92593E-31	4.044E-30
57	2850	4.81482E-32	1.011E-30
58	2900	1.20371E-32	2.528E-31
59	2950	3.00927E-33	6.319E-32
60	3000	7.52316E-34	1.58E-32
61	3050	1.88079E-34	3.95E-33
62	3100	4.70198E-35	9.874E-34
63	3150	1.17549E-35	2.469E-34
64	3200	2.93874E-36	6.171E-35
65	3250	7.34684E-37	1.543E-35
66	3300	1.83671E-37	3.857E-36

(*continued*)

Table 7. (*continued*)

Generation	Year	Birth at 50	Population
67	3350	4.59177E-38	9.643E-37
68	3400	1.14794E-38	2.411E-37
69	3450	2.86986E-39	6.027E-38
70	3500	7.17465E-40	1.507E-38
71	3550	1.79366E-40	3.767E-39
72	3600	4.48416E-41	9.417E-40
73	3650	1.12104E-41	2.354E-40
74	3700	2.8026E-42	5.885E-41
75	3750	7.00649E-43	1.471E-41
76	3800	1.75162E-43	3.678E-42
77	3850	4.37906E-44	9.196E-43
78	3900	1.09476E-44	2.299E-43
79	3950	2.73691E-45	5.748E-44
80	4000	6.84228E-46	1.437E-44

Graph 7. Life extension of 150 years, 0.5 child at age 50

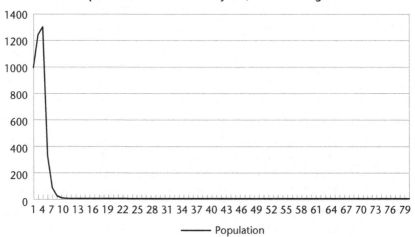

Population

Table 8.
1000-year life expectancy, 1.0 child at average age 25, 1.0 child at average age 500

Generation	Year	Birth at 25	Birth at 500	Population
0	0	0	0	1000
1	25	500	0	1500
2	50	250	0	1750
3	75	125	0	1875
4	100	62.5	0	1937.5
5	125	31.25	0	1968.75
6	150	15.625	0	1984.375
7	175	7.8125	0	1992.1875
8	200	3.90625	0	1996.09375
9	225	1.953125	0	1998.046875
10	250	0.976563	0	1999.023438
11	275	0.488281	0	1999.511719
12	300	0.244141	0	1999.755859
13	325	0.12207	0	1999.87793
14	350	0.061035	0	1999.938965
15	375	0.030518	0	1999.969482
16	400	0.015259	0	1999.984741
17	425	0.007629	0	1999.992371
18	450	0.003815	0	1999.996185
19	475	0.001907	0	1999.998093
20	500	0.000954	500	2499.999046
21	525	250.0005	250	2999.999523
22	550	250.0002	125	3374.999762
23	575	187.5001	62.5	3624.999881
24	600	125.0001	31.25	3781.24994
25	625	78.12503	15.625	3874.99997
26	650	46.87501	7.8125	3929.687485
27	675	27.34376	3.90625	3960.937493
28	700	15.625	1.953125	3978.515621
29	725	8.789064	0.9765625	3988.281248
30	750	4.882813	0.48828125	3993.652343
31	775	2.685547	0.244140625	3996.582031
32	800	1.464844	0.122070313	3998.168945

(*continued*)

Table 8. (*continued*)

Generation	Year	Birth at 25	Birth at 500	Population
33	825	0.793457	0.061035156	3999.023437
34	850	0.427246	0.030517578	3999.481201
35	875	0.228882	0.015258789	3999.725342
36	900	0.12207	0.007629395	3999.855041
37	925	0.06485	0.003814697	3999.923706
38	950	0.034332	0.001907349	3999.959946
39	975	0.01812	0.000953674	3999.979019
40	1000	0.009537	250.0004768	3249.989033
41	1025	125.005	250.0002384	3124.994278
42	1050	187.5026	187.5001192	3249.99702
43	1075	187.5014	125.0000596	3437.49845
44	1100	156.2507	78.1250298	3609.374195
45	1125	117.1879	46.8750149	3742.187083
46	1150	82.03144	27.34375745	3835.937284
47	1175	54.6876	15.62500373	3898.437388
48	1200	35.1563	8.789064363	3938.476505
49	1225	21.97268	4.882813431	3963.378876
50	1250	13.42775	2.685547341	3978.51561
51	1275	8.056648	1.464843983	3987.54882
52	1300	4.760746	0.793457148	3992.858883
53	1325	2.777102	0.427246152	3995.94116
54	1350	1.602174	0.228881865	3997.711181
55	1375	0.915528	0.122070327	3998.718261
56	1400	0.518799	0.064849861	3999.286651
57	1425	0.291824	0.034332279	3999.605179
58	1450	0.163078	0.018119814	3999.782562
59	1475	0.090599	0.009536744	3999.880791
60	1500	0.050068	125.0050068	3624.934912
61	1525	62.52754	187.5026226	3374.964595
62	1550	125.0151	187.5013709	3312.480807
63	1575	156.2582	156.2507153	3374.989629
64	1600	156.2545	117.1878725	3492.181912
65	1625	136.7212	82.03144372	3617.184497
66	1650	109.3763	54.68760058	3726.560891

(*continued*)

Table 8. (*continued*)

Generation	Year	Birth at 25	Birth at 500	Population
67	1675	82.03195	35.15630215	3812.499139
68	1700	58.59413	21.97268326	3875.487822
69	1725	40.28341	13.42774834	3919.433349
70	1750	26.85558	8.056647843	3948.974479
71	1775	17.45611	4.760745913	3968.26165
72	1800	11.10843	2.77710153	3980.560266
73	1825	6.942765	1.602173841	3988.250713
74	1850	4.27247	0.915527853	3992.980947
75	1875	2.593999	0.51879909	3995.849604
76	1900	1.556399	0.291824475	3997.568128
77	1925	0.924112	0.163078377	3998.586653
78	1950	0.543595	0.090599096	3999.184608
79	1975	0.317097	0.05006792	3999.532699
80	2000	0.183582	62.52753736	3812.233805

Graph 8. Life extension of 1000 years, 1.0 child at age 25 and 1.0 child at age 500

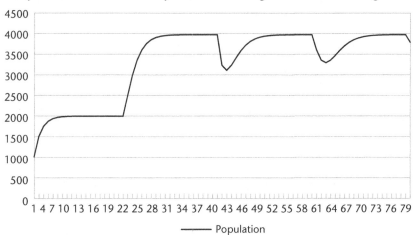

Table 9.
1000-year life expectancy, 1.0 child at average age 25, 1.0 child at average age 50, 1.0 child at average age 75

Generation	Year	Birth at 25	Birth at 50	Birth at 75	Population
0	0	1000	0	0	1000
1	25	500	0	0	1500
2	50	250	0	0	1750
3	75	125	0	0	1875
4	100	62.5	0	0	1937.5
5	125	31.25	0	0	1968.75
6	150	15.625	0	0	1984.375
7	175	7.8125	0	0	1992.1875
8	200	3.90625	0	0	1996.09375
9	225	1.953125	0	0	1998.046875
10	250	0.976563	0	0	1999.023438
11	275	0.488281	0	0	1999.511719
12	300	0.244141	0	0	1999.755859
13	325	0.12207	0	0	1999.87793
14	350	0.061035	0	0	1999.938965
15	375	0.030518	0	0	1999.969482
16	400	0.015259	0	0	1999.984741
17	425	0.007629	0	0	1999.992371
18	450	0.003815	0	0	1999.996185
19	475	0.001907	0	0	1999.998093
20	500	0.000954	500	0	2499.999046
21	525	250.0005	250	0	2999.999523
22	550	250.0002	125	0	3374.999762
23	575	187.5001	62.5	0	3624.999881
24	600	125.0001	31.25	0	3781.24994
25	625	78.12503	15.625	0	3874.99997
26	650	46.87501	7.8125	0	3929.687485
27	675	27.34376	3.90625	0	3960.937493
28	700	15.625	1.953125	0	3978.515621
29	725	8.789064	0.9765625	0	3988.281248
30	750	4.882813	0.48828125	0	3993.652343
31	775	2.685547	0.244140625	0	3996.582031

(*continued*)

Table 9. (*continued*)

Generation	Year	Birth at 25	Birth at 50	Birth at 75	Population
32	800	1.464844	0.122070313	0	3998.168945
33	825	0.793457	0.061035156	0	3999.023437
34	850	0.427246	0.030517578	0	3999.481201
35	875	0.228882	0.015258789	0	3999.725342
36	900	0.12207	0.007629395	0	3999.855041
37	925	0.06485	0.003814697	0	3999.923706
38	950	0.034332	0.001907349	0	3999.959946
39	975	0.01812	0.000953674	0	3999.979019
40	1000	0.009537	250.0004768	0	3249.989033
41	1025	125.005	250.0002384	0.009536744	3125.003815
42	1050	187.5026	187.5001192	125.0050068	3375.011563
43	1075	187.5014	125.0000596	187.5026226	3750.015616
44	1100	156.2507	78.1250298	187.5013709	4109.392732
45	1125	117.1879	46.8750149	156.2507153	4398.456335
46	1150	82.03144	27.34375745	117.1878725	4609.394409
47	1175	54.6876	15.62500373	82.03144372	4753.925957
48	1200	35.1563	8.789064363	54.68760058	4848.652674
49	1225	21.97268	4.882813431	35.15630215	4908.711348
50	1250	13.42775	2.685547341	21.97268326	4945.820764
51	1275	8.056648	1.464843983	13.42774834	4968.281723
52	1300	4.760746	0.793457148	8.056647843	4981.648433
53	1325	2.777102	0.427246152	4.760745913	4989.491457
54	1350	1.602174	0.228881865	2.77710153	4994.038579
55	1375	0.915528	0.122070327	1.602173841	4996.647833
56	1400	0.518799	0.064849861	0.915527853	4998.131751
57	1425	0.291824	0.034332279	0.51879909	4998.969078
58	1450	0.163078	0.018119814	0.291824475	4999.438286
59	1475	0.090599	0.009536744	0.163078377	4999.699592
60	1500	0.050068	125.0050068	0.090599096	4624.844313
61	1525	62.52754	187.5026226	0.05006792	4374.924064
62	1550	125.0151	187.5013709	62.52753736	4374.967813
63	1575	156.2582	156.2507153	125.01508	4562.491715
64	1600	156.2545	117.1878725	156.2582254	4835.942224
65	1625	136.7212	82.03144372	156.2544703	5117.199279

(*continued*)

Table 9. (*continued*)

Generation	Year	Birth at 25	Birth at 50	Birth at 75	Population
66	1650	109.3763	54.68760058	136.7211714	5363.296844
67	1675	82.03195	35.15630215	109.3763076	5558.6114
68	1700	58.59413	21.97268326	82.03195408	5703.632037
69	1725	40.28341	13.42774834	58.59412812	5806.171692
70	1750	26.85558	8.056647843	40.28340569	5875.996228
71	1775	17.45611	4.760745913	26.85557702	5922.138976
72	1800	11.10843	2.77710153	17.45611243	5951.893705
73	1825	6.942765	1.602173841	11.10842917	5970.692581
74	1850	4.27247	0.915527853	6.942765351	5982.36558
75	1875	2.593999	0.51879909	4.272469596	5989.506706
76	1900	1.556399	0.291824475	2.593998724	5993.819229
77	1925	0.924112	0.163078377	1.556398907	5996.394153
78	1950	0.543595	0.090599096	0.924111691	5997.916219
79	1975	0.317097	0.05006792	0.543595034	5998.807906
80	2000	0.183582	62.52753736	0.317097065	5811.826109

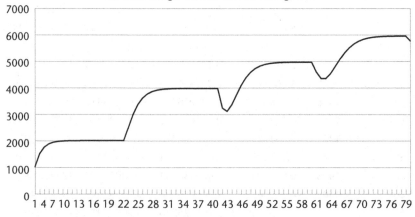

Graph 9. Life extension of 1000 years, 1.0 child at age 25,
1.0 child at age 50, and 1.0 child at age 75

Table 10.
1000-year life expectancy, 1.0 child at average age 25, 0.5 child at average age 500

Generation	Year	Birth at 25	Birth at 500	Population
0	0	0	0	1000
1	25	500	0	1500
2	50	250	0	1750
3	75	125	0	1875
4	100	62.5	0	1937.5
5	125	31.25	0	1968.75
6	150	15.625	0	1984.375
7	175	7.8125	0	1992.1875
8	200	3.90625	0	1996.09375
9	225	1.953125	0	1998.046875
10	250	0.976563	0	1999.023438
11	275	0.488281	0	1999.511719
12	300	0.244141	0	1999.755859
13	325	0.12207	0	1999.87793
14	350	0.061035	0	1999.938965
15	375	0.030518	0	1999.969482
16	400	0.015259	0	1999.984741
17	425	0.007629	0	1999.992371
18	450	0.003815	0	1999.996185
19	475	0.001907	0	1999.998093
20	500	0.000954	250	2249.999046
21	525	125.0005	125	2499.999523
22	550	125.0002	62.5	2687.499762
23	575	93.75012	31.25	2812.499881
24	600	62.50006	15.625	2890.62494
25	625	39.06253	7.8125	2937.49997
26	650	23.43751	3.90625	2964.843735
27	675	13.67188	1.953125	2980.468743
28	700	7.812504	0.9765625	2989.257809
29	725	4.394533	0.48828125	2994.140623
30	750	2.441407	0.244140625	2996.826171
31	775	1.342774	0.122070313	2998.291015
32	800	0.732422	0.061035156	2999.084472

(continued)

Table 10. (*continued*)

Generation	Year	Birth at 25	Birth at 500	Population
33	825	0.396729	0.030517578	2999.511719
34	850	0.213623	0.015258789	2999.740601
35	875	0.114441	0.007629395	2999.862671
36	900	0.061035	0.003814697	2999.927521
37	925	0.032425	0.001907349	2999.961853
38	950	0.017166	0.000953674	2999.979973
39	975	0.00906	0.000476837	2999.98951
40	1000	0.004768	62.50023842	2062.494516
41	1025	31.2525	62.50011921	1656.247139
42	1050	46.87631	46.8750596	1499.99851
43	1075	46.87569	31.2500298	1453.124225
44	1100	39.06286	19.5312649	1449.218348
45	1125	29.29706	11.71875745	1458.984166
46	1150	20.50791	6.835941225	1470.703017
47	1175	13.67193	3.906251863	1480.468694
48	1200	8.789089	2.197266556	1487.548799
49	1225	5.493178	1.220703591	1492.309555
50	1250	3.356941	0.671386952	1495.36132
51	1275	2.014164	0.366211054	1497.253414
52	1300	1.190187	0.198364316	1498.397825
53	1325	0.694276	0.106811553	1499.076842
54	1350	0.400544	0.057220474	1499.473571
55	1375	0.228882	0.030517585	1499.702453
56	1400	0.1297	0.016212467	1499.833107
57	1425	0.072956	0.008583071	1499.907017
58	1450	0.04077	0.004529954	1499.948502
59	1475	0.02265	0.002384186	1499.971628
60	1500	0.012517	15.6262517	1265.609443
61	1525	7.819384	23.43815565	1046.866506
62	1550	15.62877	23.43784273	898.4328806
63	1575	19.53331	19.53142881	812.4974966
64	1600	19.53237	14.64853063	768.5533352
65	1625	17.09045	10.25395468	749.0227092
66	1650	13.6722	6.835962646	742.1871088

(*continued*)

Table 10. (*continued*)

Generation	Year	Birth at 25	Birth at 500	Population
67	1675	10.25408	4.394544289	741.210728
68	1700	7.324313	2.746588783	742.4925638
69	1725	5.035451	1.678470289	744.3236707
70	1750	3.356961	1.007081883	746.0021655
71	1775	2.182021	0.595093705	747.3144362
72	1800	1.388557	0.347137931	748.2566744
73	1825	0.867848	0.200271854	748.8975478
74	1850	0.53406	0.114441045	749.3171667
75	1875	0.32425	0.064849919	749.5841967
76	1900	0.19455	0.036478076	749.7503751
77	1925	0.115514	0.020384806	749.8519417
78	1950	0.067949	0.011324891	749.9130962
79	1975	0.039637	0.006258492	749.9494552
80	2000	0.022948	3.909692171	691.3770884

Graph 10. Life extension of 1000 years, 1.0 child at age 25 and 0.5 child at age 500

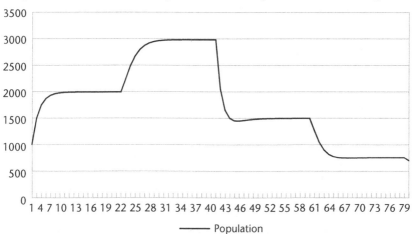

Population

Table 11.

1000-year life expectancy, 1.0 child at average age 25

Generation	Year	Birth at 25	Population
0	0	0	1000
1	25	500	1500
2	50	250	1750
3	75	125	1875
4	100	62.5	1937.5
5	125	31.25	1968.75
6	150	15.625	1984.375
7	175	7.8125	1992.188
8	200	3.90625	1996.094
9	225	1.953125	1998.047
10	250	0.9765625	1999.023
11	275	0.48828125	1999.512
12	300	0.244140625	1999.756
13	325	0.122070313	1999.878
14	350	0.061035156	1999.939
15	375	0.030517578	1999.969
16	400	0.015258789	1999.985
17	425	0.007629395	1999.992
18	450	0.003814697	1999.996
19	475	0.001907349	1999.998
20	500	0.000953674	1999.999
21	525	0.000476837	2000
22	550	0.000238419	2000
23	575	0.000119209	2000
24	600	5.96046E-05	2000
25	625	2.98023E-05	2000
26	650	1.49012E-05	2000
27	675	7.45058E-06	2000
28	700	3.72529E-06	2000
29	725	1.86265E-06	2000
30	750	9.31323E-07	2000
31	775	4.65661E-07	2000
32	800	2.32831E-07	2000

(*continued*)

Table 11. (*continued*)

Generation	Year	Birth at 25	Population
33	825	1.16415E-07	2000
34	850	5.82077E-08	2000
35	875	2.91038E-08	2000
36	900	1.45519E-08	2000
37	925	7.27596E-09	2000
38	950	3.63798E-09	2000
39	975	1.81899E-09	2000
40	1000	9.09495E-10	1000
41	1025	4.54747E-10	500
42	1050	2.27374E-10	250
43	1075	1.13687E-10	125
44	1100	5.68434E-11	62.5
45	1125	2.84217E-11	31.25
46	1150	1.42109E-11	15.625
47	1175	7.10543E-12	7.8125
48	1200	3.55271E-12	3.90625
49	1225	1.77636E-12	1.953125
50	1250	8.88178E-13	0.976562
51	1275	4.44089E-13	0.488281
52	1300	2.22045E-13	0.244141
53	1325	1.11022E-13	0.12207
54	1350	5.55112E-14	0.061035
55	1375	2.77556E-14	0.030518
56	1400	1.38778E-14	0.015259
57	1425	6.93889E-15	0.007629
58	1450	3.46945E-15	0.003815
59	1475	1.73472E-15	0.001907
60	1500	8.67362E-16	0.000954
61	1525	4.33681E-16	0.000477
62	1550	2.1684E-16	0.000238
63	1575	1.0842E-16	0.000119
64	1600	5.42101E-17	5.96E-05
65	1625	2.71051E-17	2.98E-05
66	1650	1.35525E-17	1.49E-05

(*continued*)

Table 11. (*continued*)

Generation	Year	Birth at 25	Population
67	1675	6.77626E-18	7.45E-06
68	1700	3.38813E-18	3.73E-06
69	1725	1.69407E-18	1.86E-06
70	1750	8.47033E-19	9.31E-07
71	1775	4.23516E-19	4.66E-07
72	1800	2.11758E-19	2.33E-07
73	1825	1.05879E-19	1.16E-07
74	1850	5.29396E-20	5.82E-08
75	1875	2.64698E-20	2.91E-08
76	1900	1.32349E-20	1.46E-08
77	1925	6.61744E-21	7.28E-09
78	1950	3.30872E-21	3.64E-09
79	1975	1.65436E-21	1.82E-09
80	2000	8.27181E-22	9.09E-10

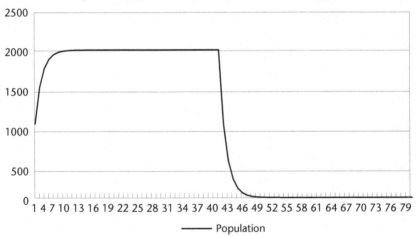

Graph 11. Life extension of 1000 years, 1.0 child at age 25

Table 12.
1000-year life expectancy, 0.5 child at average age 25

Generation	Year	Birth at 25	Population
0	0	0	1000
1	25	250	1250
2	50	62.5	1312.5
3	75	15.625	1328.125
4	100	3.90625	1332.031
5	125	0.9765625	1333.008
6	150	0.244140625	1333.252
7	175	0.061035156	1333.313
8	200	0.015258789	1333.328
9	225	0.003814697	1333.332
10	250	0.000953674	1333.333
11	275	0.000238419	1333.333
12	300	5.96046E-05	1333.333
13	325	1.49012E-05	1333.333
14	350	3.72529E-06	1333.333
15	375	9.31323E-07	1333.333
16	400	2.32831E-07	1333.333
17	425	5.82077E-08	1333.333
18	450	1.45519E-08	1333.333
19	475	3.63798E-09	1333.333
20	500	9.09495E-10	1333.333
21	525	2.27374E-10	1333.333
22	550	5.68434E-11	1333.333
23	575	1.42109E-11	1333.333
24	600	3.55271E-12	1333.333
25	625	8.88178E-13	1333.333
26	650	2.22045E-13	1333.333
27	675	5.55112E-14	1333.333
28	700	1.38778E-14	1333.333
29	725	3.46945E-15	1333.333
30	750	8.67362E-16	1333.333
31	775	2.1684E-16	1333.333
32	800	5.42101E-17	1333.333

(continued)

Table 12. (*continued*)

Generation	Year	Birth at 25	Population
33	825	1.35525E-17	1333.333
34	850	3.38813E-18	1333.333
35	875	8.47033E-19	1333.333
36	900	2.11758E-19	1333.333
37	925	5.29396E-20	1333.333
38	950	1.32349E-20	1333.333
39	975	3.30872E-21	1333.333
40	1000	8.27181E-22	333.3333
41	1025	2.06795E-22	83.33333
42	1050	5.16988E-23	20.83333
43	1075	1.29247E-23	5.208333
44	1100	3.23117E-24	1.302083
45	1125	8.07794E-25	0.325521
46	1150	2.01948E-25	0.08138
47	1175	5.04871E-26	0.020345
48	1200	1.26218E-26	0.005086
49	1225	3.15544E-27	0.001272
50	1250	7.88861E-28	0.000318
51	1275	1.97215E-28	7.95E-05
52	1300	4.93038E-29	1.99E-05
53	1325	1.2326E-29	4.97E-06
54	1350	3.08149E-30	1.24E-06
55	1375	7.70372E-31	3.1E-07
56	1400	1.92593E-31	7.76E-08
57	1425	4.81482E-32	1.94E-08
58	1450	1.20371E-32	4.85E-09
59	1475	3.00927E-33	1.21E-09
60	1500	7.52316E-34	3.03E-10
61	1525	1.88079E-34	7.57E-11
62	1550	4.70198E-35	1.89E-11
63	1575	1.17549E-35	4.66E-12
64	1600	2.93874E-36	1.11E-12
65	1625	7.34684E-37	2.2E-13
66	1650	1.83671E-37	-1.8E-15

(*continued*)

Table 12. (*continued*)

Generation	Year	Birth at 25	Population
67	1675	4.59177E-38	-5.7E-14
68	1700	1.14794E-38	-7.1E-14
69	1725	2.86986E-39	-7.5E-14
70	1750	7.17465E-40	-7.6E-14
71	1775	1.79366E-40	-7.6E-14
72	1800	4.48416E-41	-7.6E-14
73	1825	1.12104E-41	-7.6E-14
74	1850	2.8026E-42	-7.6E-14
75	1875	7.00649E-43	-7.6E-14
76	1900	1.75162E-43	-7.6E-14
77	1925	4.37906E-44	-7.6E-14
78	1950	1.09476E-44	-7.6E-14
79	1975	2.73691E-45	-7.6E-14
80	2000	6.84228E-46	-7.6E-14

Graph 12. Life extension of 1000 years, 0.5 child at age 25

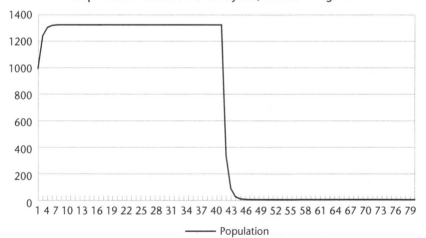

—— Population

Table 13.
1000-year life expectancy, 0.5 child at average age 50

Generation	Year	Birth at 50	Population
0	0	0	1000
1	50	250	1250
2	100	62.5	1312.5
3	150	15.625	1328.125
4	200	3.90625	1332.0313
5	250	0.9765625	1333.0078
6	300	0.244140625	1333.252
7	350	0.061035156	1333.313
8	400	0.015258789	1333.3282
9	450	0.003814697	1333.3321
10	500	0.000953674	1333.333
11	550	0.000238419	1333.3333
12	600	5.96046E-05	1333.3333
13	650	1.49012E-05	1333.3333
14	700	3.72529E-06	1333.3333
15	750	9.31323E-07	1333.3333
16	800	2.32831E-07	1333.3333
17	850	5.82077E-08	1333.3333
18	900	1.45519E-08	1333.3333
19	950	3.63798E-09	1333.3333
20	1000	9.09495E-10	333.33333
21	1050	2.27374E-10	83.333333
22	1100	5.68434E-11	20.833333
23	1150	1.42109E-11	5.2083333
24	1200	3.55271E-12	1.3020833
25	1250	8.88178E-13	0.3255208
26	1300	2.22045E-13	0.0813802
27	1350	5.55112E-14	0.0203451
28	1400	1.38778E-14	0.0050863
29	1450	3.46945E-15	0.0012716
30	1500	8.67362E-16	0.0003179
31	1550	2.1684E-16	7.947E-05
32	1600	5.42101E-17	1.987E-05

(continued)

Table 13. (*continued*)

Generation	Year	Birth at 50	Population
33	1650	1.35525E-17	4.967E-06
34	1700	3.38813E-18	1.242E-06
35	1750	8.47033E-19	3.104E-07
36	1800	2.11758E-19	7.761E-08
37	1850	5.29396E-20	1.94E-08
38	1900	1.32349E-20	4.851E-09
39	1950	3.30872E-21	1.213E-09
40	2000	8.27181E-22	3.032E-10
41	2050	2.06795E-22	7.579E-11
42	2100	5.16988E-23	1.895E-11
43	2150	1.29247E-23	4.737E-12
44	2200	3.23117E-24	1.184E-12
45	2250	8.07794E-25	2.961E-13
46	2300	2.01948E-25	7.401E-14
47	2350	5.04871E-26	1.85E-14
48	2400	1.26218E-26	4.626E-15
49	2450	3.15544E-27	1.156E-15
50	2500	7.88861E-28	2.891E-16
51	2550	1.97215E-28	7.228E-17
52	2600	4.93038E-29	1.807E-17
53	2650	1.2326E-29	4.518E-18
54	2700	3.08149E-30	1.129E-18
55	2750	7.70372E-31	2.823E-19
56	2800	1.92593E-31	7.059E-20
57	2850	4.81482E-32	1.765E-20
58	2900	1.20371E-32	4.412E-21
59	2950	3.00927E-33	1.103E-21
60	3000	7.52316E-34	2.757E-22
61	3050	1.88079E-34	6.893E-23
62	3100	4.70198E-35	1.723E-23
63	3150	1.17549E-35	4.308E-24
64	3200	2.93874E-36	1.077E-24
65	3250	7.34684E-37	2.693E-25
66	3300	1.83671E-37	6.732E-26

(*continued*)

Table 13. (*continued*)

Generation	Year	Birth at 50	Population
67	3350	4.59177E-38	1.683E-26
68	3400	1.14794E-38	4.207E-27
69	3450	2.86986E-39	1.052E-27
70	3500	7.17465E-40	2.63E-28
71	3550	1.79366E-40	6.574E-29
72	3600	4.48416E-41	1.643E-29
73	3650	1.12104E-41	4.109E-30
74	3700	2.8026E-42	1.027E-30
75	3750	7.00649E-43	2.568E-31
76	3800	1.75162E-43	6.42E-32
77	3850	4.37906E-44	1.605E-32
78	3900	1.09476E-44	4.012E-33
79	3950	2.73691E-45	1.003E-33
80	4000	6.84228E-46	2.508E-34

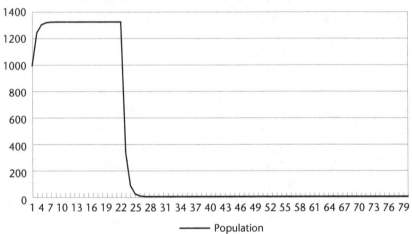

Graph 13. Life extension of 1000 years, 0.5 child at age 50

Notes

Chapter 1

1. Natalie Elliot, "The Politics of Life Extension in Francis Bacon's *Wisdom of the Ancients*," *Review of Politics* 77 (2015): 351–375.

2. Leonard Hayflick, *How and Why We Age* (New York: Ballantine Books, 1994), 267–271.

3. The term *geroscience* was coined a few years ago by scientists at the National Institutes of Health. *Biogerontology* is the traditional term for the science of aging.

4. Richard Miller, "Extending Life: Scientific Prospects and Political Obstacles," *Milbank Quarterly* 80, no. 1 (2002): 166. See also Paola S. Timiras, "Comparative and Differential Aging, Geriatric Functional Assessment, Aging and Disease," in *Physiological Basis of Aging and Geriatrics*, ed. Paola S. Timiras, 3rd ed. (New York: CRC Press, 2003), 25–46.

5. Their optimism is also shared by a large group of technology enthusiasts who call themselves "transhumanists" and believe we may one day halt aging using highly futuristic methods, such as putting nanomachines in our bloodstream to repair our molecules and cells or uploading our minds into computers. See David Gelles, "Immortality 2.0: A Silicon Valley Insider Looks at California's Transhumanist Movement," *Futurist* 43, no. 1 (2009): 34–41. These predictions go beyond anything that mainstream scientists foresee and what I will discuss here.

6. Quoted in David Stipp, "The Hunt for the Youth Pill," *Fortune* 140, no. 7 (1999): 199–216.

7. S. Jay Olshansky et al., "Position Statement on Human Aging," *Journals of Gerontology Series A: Biological Sciences and Medical Sciences* 57 (2002): B292–B297.

8. George M. Martin et al., "Research on Aging: The End of the Beginning," *Science* 299 (February 28, 2003): 1339–1341.

9. Alice Park, "Age Disrupters: A Drug from Dirt and Some Siamese Mice Have Researchers Inching toward the Seemingly Impossible: A Cure for Aging," *Time* 185, nos. 6–7 (February 23, 2015–March 2, 2015): 73–76, 74.

10. Matt Kaeberlein, Peter S. Rabinovitch, and George M. Martin, "Healthy Aging: The Ultimate Preventative Medicine," *Science* 350, no. 6265 (December 4, 2015): 1191–1193, 1191.

11. Jon Cohen, "Death-Defying Experiments: Pushing the Limits of Life Span in Animals Could Someday Help Lengthen Our Own," *Science* 350, no. 6265 (December 4, 2015): 1186.

12. Apoptosis occurs when a cell commits suicide through a genetically regulated process; cells that detect excessive damage to their DNA or have turned cancerous may commit suicide to remove themselves from the organism. If they sometimes do this when they should not, then apoptosis may be a basic process of aging.

13. Elizabeth H. Blackburn, Elissa S. Epel, and Jue Lin, "Human Telomere Biology: A Contributory and Interactive Factor in Aging, Disease Risks, and Protection," *Science* 350, no. 6265 (December 4, 2015): 1193–1198.

14. Richard Weindruch, "Caloric Restriction and Aging," *Scientific American* 274, no. 1 (1996): 46–52.

15. Some scientists suggest that evolution selects for this ability so that populations will survive famine conditions and resume breeding later (if they reproduce during the famine, their offspring are less likely to survive). John R. Speakman and Sharon E. Mitchell, "Caloric Restriction," *Molecular Aspects of Medicine* 32, no. 3 (2011): 159–221, http://www.sciencedirect.com/science/article/pii/S009829971100032X (accessed on May 6, 2017).

16. In 2012, the press widely reported that a study showed caloric restriction did not work in monkeys, but this may be an overstatement. There are two studies, one at the Wisconsin National Primate Research Center, which concluded in 2009 that the diet slows aging in monkeys, and another at the National Institute of Aging, which concluded in 2012 that it does not. University of Wisconsin-Madison News, http://news.wisc.edu/calorie-restriction-lets-monkeys-live-long-and-prosper/ (accessed on May 6, 2017); National Institute of Aging, "NIH Study Finds Calorie Restriction Does Not Affect Survival," https://www.nia.nih.gov/news/nih-study-finds-calorie-restriction-does-not-affect-survival (accessed on May 6, 2017).

The Wisconsin study has been criticized for putting its control group monkeys on an unhealthy diet, which may have produced misleading results. However, we already know that the caloric restriction diet does not work in all species; it does not work in all strains of mice, for example. Thus, the monkey studies do not prove it cannot work in some species of monkeys or in other primates—or in us.

17. Kaeberlein, Rabinovitch, and Martin, "Healthy Aging," 1192.

18. Phyllis K. Stein et al., "Caloric Restriction May Reverse Age-Related Autonomic Decline in Humans," *Aging Cell* 11, no. 4 (May 21, 2012): 644–650, http://onlinelibrary.wiley.com/doi/10.1111/j.1474-9726.2012.00825.x/full (accessed on May 7, 2017).

19. Mark A. Lane, Donald K. Ingram, and George S. Roth, "The Serious Search for an Anti-aging Pill," *Scientific American* 287, no. 2 (August 2002): 36–41.

20. Konstantinos Sousounis, Joelle A. Baddour, and Panagiotis A. Tsonis, "Aging and Regeneration in Vertebrates," *Current Topics in Developmental Biology* 108 (2014): 217–246; Y. Peng et al., "Stem Cells and Aberrant Signaling of Molecular Systems in Skin Aging," *Ageing Research Review* 19 (January 2015): 8–21; So-ichiro Fukada, Yuran Ma, and Akiyoshi Uezumi, "Adult Stem Cell and Mesenchymal Progenitor Theories of Aging," *Frontiers in Cell and Developmental Biology* 2, no. 10 (March 28, 2014): 1–9.

21. Siegfried Hekimi and Leonard Guarente, "Genetics and the Specificity of the Aging Process," *Science* 299, no. 5611 (2003): 1351–1354; George L. Sutphin and Matt Kaeberlein, "Comparative Genetics of Aging," in *Handbook of the Biology of Aging*, ed. Edward J. Masoro and Steven N. Austad, 7th ed. (London: Academic Press, 2011), 216; and Daniel E. L. Promislow, Kenneth M. Fedorka, and Joep M. S. Burger, "Evolutionary Biology of Aging: Future Directions," in *Handbook of the Biology of Aging*, ed. Masoro and Austad, 218, 221, 234.

22. Stephen N. Austad, *Why We Age: What Science Is Discovering about the Body's Journey through Life* (New York: John Wiley & Sons, 1997), 219–221; William R. Clark, *A Means to an End: The Biological Basis of Aging and Death* (Oxford: Oxford University Press, 2002), 217.

23. Sutphin and Kaeberlein, "Comparative Genetics of Aging," 231.

24. Park, "Age Disrupters," 75–76.

25. Kaeberlein, Rabinovitch, and Martin, 1192.

26. Bill Gifford, "Does a Real Anti-aging Pill Already Exist? Inside Novartis's Push to Produce the First Legitimate Anti-aging Drug," *Bloomberg*, February 12, 2015, https://www.bloomberg.com/news/features/2015-02-12/does-a-real-anti-aging-pill-already-exist- (accessed on May 3, 2017).

27. Bryan Warner, "Metformin: Long Lifespan in a Pill?," *Healthy Balance MD*, December 15, 2015, http://www.healthybalancemd.com/metformin-longer-lifespan-pill/ (accessed on May 2, 2017).

28. Kaeberlein, Rabinovitch, and Martin, "Healthy Aging," 1192.

29. Park, "Age Disrupters," 75.

30. Kaeberlein, Rabinovitch, and Martin, "Healthy Aging," 1192.

31. Kaeberlein, Rabinovitch, and Martin, "Healthy Aging," 1192. For a slightly more skeptical review of this strategy, see Ying Wang and Siegfried Hekimi, "Mitochondrial Dysfunction and Longevity in Animals: Untangling the Knot," *Science* 350, no. 6265 (December 4, 2015): 1204–1207.

32. Eric Verdin, "NAD in Aging, Metabolism, and Neurodegeneration," *Science* 350, no. 6265 (December 4, 2015): 1213.

33. Bec Crew, "Japanese Scientists Reverse Ageing in Human Cell Lines: They Made a 97-Year-Old Cell Line Behave as Good as New," *Science Alert*, May 28, 2015, http://www.sciencealert.com/japanese-scientists-reverse-ageing-in-human-cell-lines (accessed on May 2, 2017).

34. Bob Yirka, "Researchers Find a Link between Age Related Memory Loss and a Specific Enzyme in Mice," *Medical Press*, July 2, 2012, http://medicalxpress.com/news/2012-07-link-age-memory-loss-specific.html (accessed on May 2, 2017).

35. "How Aging Impairs Immune Response," *ScienceDaily*, July 12, 2012, http://www.sciencedaily.com/releases/2012/07/120717162133.htm (accessed on May 2, 2017).

36. Christine Dell'Amore, "Old Mice Made 'Young'—May Lead to Anti-aging Treatments," *National Geographic*, January 7, 2012, http://news.nationalgeographic.com/news/2012/01/120106-aging-mice-stem-cells-old-young-science-health/ (accessed on May 2, 2017).

37. Kaeberlein, Rabinovitch, and Martin, "Healthy Aging," 1192.

38. Kaeberlein, Rabinovitch, and Martin, 1191–1192. See also Emily Underwood, "The Final Countdown: In the Race to Find a Biological Clock, There Are Plenty of Contenders," *Science* 350, no. 6265 (December 4, 2015): 1188–1190.

39. Kaeberlein, Rabinovitch, and Martin, "Healthy Aging," 1192.

40. Brian M. Delaney and Lisa Walford, *The Longevity Diet: The Only Proven Way to Slow the Aging Process and Maintain Peak Vitality through Caloric Restriction*, 3rd ed. (Boston: Da Capo Lifelong Books, 2010). Search for "caloric restriction" in Amazon.com and you'll find several others.

41. Available online for free at Chronometer, https://cronometer.com/download/ (accessed on May 2, 2017).

42. David Stipp, "Beyond Resveratrol: The Anti-aging NAD Fad," *Scientific American Guest Blog*, March 11, 2015, https://blogs.scientificamerican.com/guest-blog/beyond-resveratrol-the-anti-aging-nad-fad/ (accessed on May 6, 2017).

43. Gifford, "Real Anti-aging Pill."

44. Austad and S. Jay Olshansky have a $150 bet that someone alive today will still be alive in 2150. The money has been invested; the proceeds will be paid to Austad's descendants if there is such a person and to Olshansky's if there is not. Karen Wright and Mary Ellen Mark, "Staying Alive," *Discover*, November 6, 2003, http://discovermagazine.com/2003/nov/cover (accessed on May 2, 2017).

45. Miller, "Extending Life."

46. S. Jay Olshansky, "Replacement Parts: A Survey of Recent Articles," *Wilson Quarterly* 26 (Summer 2002): 102.

47. Quoted in Constance Holden, "The Quest to Reverse Time's Toll," *Science* 295, no. 5557 (February 8, 2002): 1032.

48. Stephen S. Hall, "Kenyon's Ageless Quest," *Smithsonian*, March 2004, http://www.smithsonianmag.com/science-nature/kenyons-ageless-quest-103525532/?all (accessed on May 2, 2017).

49. Theo Richel, "Will Human Life Expectancy Quadruple in the Next Hundred Years? Sixty Gerontologists Say Public Debate of Life Extension Is Necessary," *Journal of Anti-aging Medicine* 6, no. 4 (2003): 309–314.

50. Austad, *Why We Age*, 10.

51. Simone de Beauvoir, *All Men Are Mortal*, trans. Leonard M. Friedman (New York: W. W. Norton, 1992).

52. Centers for Disease Control and Prevention, "Leading Causes of Death," http://www.cdc.gov/nchs/fastats/leading-causes-of-death.htm (accessed on May 2, 2017).

53. Pew Research Center, "Living to 120 and Beyond," http://www.pewforum.org/2013/08/06/living-to-120-and-beyond/ (accessed on May 2, 2017).

54. Nick Dragojlovic, "Canadians' Support for Radical Life Extension Resulting from Advances in Regenerative Medicine," *Journal of Aging Studies* 27, no. 2 (2013): 151–158.

55. Brad Partridge et al., "Public Attitudes towards Human Life Extension by Intervening in Ageing," *Journal of Aging Studies* 25, no. 2 (2011): 73–83. See also Allen Alvarez, Lumberto Mendoza, and Peter Danielson, "Mixed Views about Radical Life Extension," *Etikki i praksis—Nordic Journal of Applied Ethics* 9, no. 1 (2015): 87–110, a survey of 326 university students. Of that group, 13 percent agreed that life extension research is desirable, 50 percent were opposed, and 37 percent were neutral on the question.

56. Donald W. Bruckner, "In Defense of Adaptive Preferences," *Philosophical Studies* 142, no. 3 (2009): 308.

57. Brad Partridge et al., "Ethical Concerns in the Community about Technologies to Extend Human Life Span," *American Journal of Bioethics* 9, no. 12 (2009): 68–76.

58. Partridge et al., 68–76.

59. Partridge et al., 71.

60. Pew Research Center, "Living to 120 and Beyond: Americans' Views on Aging, Medical Advances and Radical Life Extension," http://www.pewforum.org/2013/08/06/living-to-120-and-beyond-americans-views-on-aging-medical-advances-and-radical-life-extension/ (accessed on May 2, 2017).

61. Mitch Leslie, "The Secret of Antiaging Is Worth Looking For," *Science* 375, no. 6350 (October 8, 2013): http://news.sciencemag.org/economics/2013/10/secret-antiaging-worth-looking (accessed on May 2, 2017); S. Jay Olshansky et al., "In Pursuit of the Longevity Dividend," *Scientist*, March 1, 2006, 28–36, http://www.the-scientist.com/?articles.view/articleNo/23784/title/The-Longevity-Dividend/ (accessed on May 2, 2017).

62. Olshansky, "The Longevity Dividend."

63. National Institutes of Health, "Estimates of Funding for Various Research, Condition, and Disease Categories (RCDC)," http://report.nih.gov/categorical_spending.aspx (accessed on May 2, 2017).

64. National Institutes of Health, "Estimates of Funding."

65. National Institute of Aging, "Fiscal Year 2017 Budget," https://www.nia.nih .gov/about/budget/2016/fy-2017-justification-budget-request/fy-2017-program -descriptions-and (accessed on May 2, 2017).

66. The Ellison Medical Foundation, founded by Lawrence J. Ellison, founder and CEO of Oracle, is a nonprofit organization focused on basic biomedical research on slowing aging and how to prevent age-related diseases. Lawrence Ellison Foundation, http://www.ellisonfoundation.org/ (accessed on May 2, 2017). The current webpage does not disclose how much this foundation has awarded to date, but a now-defunct webpage accessed in 2015 said that it has awarded nearly $45 million in grants since the late 1990s.

 The Methuselah Foundation, cofounded by entrepreneurs David Gobel and Aubrey de Grey, is also focused on aging and age-related disease: it has awarded over $4 million for research and development in regenerative medicine. Methuselah Foundation, https://mfoundation.org/ (accessed on May 2, 2017).

Some money from the Methuselah Foundation is routed to the SENS Research Foundation, headed by Aubrey de Grey, which conducts and funds research on regenerative medicine and distributed $2,797,124 for antiaging research in 2014 alone. SENS Research Foundation, "Annual Report," http://www.sens.org/sites/srf .org/files/reports/SENS%20Research%20Foundation%20Annual%20Report%202015 .pdf (accessed on May 2, 2017).

Other foundations working in this area include the Glenn Foundation, "Glenn Foundation for Medical Research," http://www.glennfoundation.org/ (accessed on May 2, 2017); and the Buck Institute, "Buck Institute for Research on Aging," http:// www.buckinstitute.org/ (accessed on May 2, 2017).

67. Tad Friend, "Silicon Valley's Quest to Live Forever," *New Yorker*, April 3, 2017, http://www.newyorker.com/magazine/2017/04/03/silicon-valleys-quest-to-live -forever (accessed on May 5, 2017).

68. Friend, "Silicon Valley's Quest."

69. Calico, http://www.calicolabs.com/ (accessed on May 2, 2017). See also Arion McNicoll, "How Google's Calico Aims to Fight Aging and 'Solve Death,'" *CNNTech*, October 3, 2013, http://www.cnn.com/2013/10/03/tech/innovation/google-calico -aging-death/ (accessed on May 2, 2017); and Friend, "Silicon Valley's Quest."

70. News release, Calico, http://www.calicolabs.com/news/2014/09/03/ (accessed on May 2, 2017).

Chapter 2

1. Partridge et al., "Ethical Concerns," 71.

2. Leon R. Kass, "L'Chaim and Its Limits: Why Not Immortality?," in *The Fountain of Youth: Cultural, Scientific, and Ethical Perspectives on a Biomedical Goal*, ed. Stephen G.

Post and Robert H. Binstock (Oxford: Oxford University Press, 2004), 312; Daniel Callahan, *False Hopes: Why America's Quest for Perfect Health Is a Recipe for Failure* (New York: Simon and Schuster, 1998), 132.

3. Bernard Williams, "The Makropulos Case: Reflections on the Tedium of Immortality," in *The Metaphysics of Death*, ed. John Martin Fischer (Stanford: Stanford University Press, 1993), 73–92.

4. Williams, "The Makropulos Case," 73, 81–83.

5. Williams actually considers three possibilities, so he presents us with a trilemma, not a dilemma. The third possibility is just like the second, with one difference: you change just as much as you do in the second possibility, but you don't cease to be the same person. In other words, you change just as much, but you still exist. You still exist not because there is something different about you in the third possibility but because the correct theory of personal identity turns out to be different than we thought it was. What other theories of personal identity are there? Perhaps you exist as long as your soul exists (where your soul is different from your character and personality), or perhaps you exist as long as your physical brain exists (allowing for continual substitution of new molecules for old ones), or something like that. Such theories are not popular among philosophers, and Williams seems skeptical of them too. However, his point is not that such theories might be true. Rather, his point is that simply being identical with some future self does not, by itself, give you a reason to want that future self to exist. His reasons for believing this are complex and subtle, and I discuss the third possibility in appendix B. There I argue that his claims about the third possibility run into the same problems as his claims about changing personal identity. Since the problems are the same, I will not repeat that material here.

6. Kevin Aho, "Simmel on Acceleration, Boredom, and Extreme Aesthesia," *Journal for the Theory of Social Behavior* 37, no. 4 (2007): 447.

7. Lisa Bortolotti and Yujin Nagasawa, "Immortality without Boredom," *Ratio* 22, no. 3 (September 2009): 268–271.

8. R. Bargdill, "The Study of Life Boredom," *Journal of Phenomenological Psychology* 31 (2000): 188–219.

9. Bortolotti and Nagasawa, "Immortality without Boredom," 268.

10. John Martin Fischer, "Immortality," in *The Oxford Handbook of Philosophy of Death*, ed. Ben Bradley, Fred Feldman, and Jens Johansson (Oxford: Oxford University Press, 2012), 336–354. See also Timothy Chappell, "Infinity Goes Up On Trial: Must Immortality Be Meaningless?," *European Journal of Philosophy* 17, no. 1 (2009): 42.

11. Williams, "The Makropulos Case," 87.

12. Robert Nozick, *Anarchy, State, and Utopia* (New York: Basic Books, 1974), 42–45.

13. Larry Temkin, "Is Living Longer Living Better?," *Journal of Applied Philosophy* 25, no. 3 (August 2008): 193–210.

14. Walter Glannon, "Identity, Prudential Concern, and Extended Lives," *Bioethics* 16, no. 3 (2002): 268.

15. Many of them also say that there must be *psychological connectedness* between you and all your earlier selves. Psychological connectedness can take the form of similarity; for example, you might share a lot of memories and other psychological traits with your earliest self. However, psychological connectedness does not have to take the form of similarity. It can also take the form of causal connections between your earlier and later psychological states: the later ones exist because of the earlier ones (in some sense). Therefore, even personal identity theories that require connectedness do not thereby require similarity, even though connectedness may well take the form of similarity.

16. John Harris, "A Response to Walter Glannon," *Bioethics* 16, no. 3 (2002): 286; see also Chappell, "Infinity Goes On Trial," 39. Some writers have also argued that if S2 (your self 40 years from now) has a concern for S3 (your self 80 years from now), then S1 (you now) cares about S3 indirectly because S1 wants S2's desires to be fulfilled, and S3 cares about S4's desires, so S1 cares about S4 indirectly too, and so on. (See John Schloendorn, "Making the Case for Human Life Extension: Personal Arguments," *Bioethics* 20, no. 4 [2006]; and Chappell, "Infinity Goes On Trial," 39.) This is a variant of the idea that you might care that all your future categorical desires—whatever they may be—will be fulfilled. (For more on categorical desires, see appendix B.)

I am not sure this works without something like the Tarzan condition. As I discussed in section 2.2, Williams is interested in whether an extremely long life span is good for you, not whether it is good for someone else you care about. From S1 to S7 (let's say), there are no shared categorical desires except a chain of desires that the next stage flourish. S2 had no desire that S7 flourish; neither did S3. To say that S1 indirectly cares about S7 is simply to say that S1 desires that S7 flourish and that his categorical desires be fulfilled. But S1 doesn't desire that for S7; he desires it for S2.

17. This is not necessarily immortality; you could be alive forever and still be capable of dying. Let us say that immortality is *hard* if you cannot die and *soft* if you can die but, as it happens, you never do (granted the odds of this are vanishingly small, but the distinction is real). You might want it either way: you might want hard immortality, or you might now want to be alive forever but nonetheless hedge your bets against a future change of mind by retaining the option of death if life becomes intolerable, and thereby want soft immortality.

18. This might overlap with the first way; at each time in the future, you might want hard immortality—and already have it. This happens if you desire hard immortality, get it, and remain forever glad you've got it.

19. Either of these desires can be reason to desire immortality, and either can be reason to desire extended life. The one catch is that if you get extended life, a desire to exist at all times in the future will eventually be frustrated, for you will die

eventually from something or other. Even so, extended life will get you closer to existing at all times in the future than a normal life span would, so it is worth seeking whether you have an eternity desire or an ongoing desire.

20. Christine Overall argues that you might rationally want to change your character, personality, and values dramatically. On her view, the change would not be a bad thing; it would be a *good* thing and a reason in favor of life extension. *Aging, Death, and Human Longevity: A Philosophical Inquiry* (Berkeley: University of California Press, 2003), 158–159.

Chapter 3

1. Portions of this chapter, particularly section 3.6, also appear in John K. Davis, "Four Ways Life Extension Will Change Our Relationship with Death," *Bioethics* 30, no. 3 (2016): 165–172.

2. Leon R. Kass, "L'Chaim and Its Limits: Why Not Immortality?," in *Fountain of Youth*, ed. Post and Binstock, 311. Kass is well-known for his opposition to life extension, though nearly everything he says appears in one essay, first published as Leon R. Kass, "L'Chaim and Its Limits: Why Not Immortality?," *First Things* 113 (May 2001); it has since been published several places under various titles, with minor modifications, including Leon R. Kass, "The Case for Mortality," *American Scholar* 52, no. 2 (Spring 1983): 173–191. See also Leon R. Kass, "L'Chaim and Its Limits: Why Not Immortality?," chap. 9 in *Life, Liberty and the Defense of Dignity: The Challenge for Bioethics* (New York: Encounter Books, 2002); and President's Council on Bioethics, "Ageless Bodies," chap. 4 in *Beyond Therapy*, http://bioethics.georgetown.edu/pcbe/reports/beyondtherapy/ (accessed on May 6, 2017). From here on, all citations to this article will refer to the *Fountain of Youth* version.

In "L'Chaim," Kass offers four arguments that extended life is not good for you: living within a normal life span (1) prepares us to accept death, (2) forces us to use our time on this earth well and make our lives count for something, (3) deepens our character by forcing us to give up attachment to personal survival, and (4) enables us to avoid boredom by becoming engaged with and serious about life. I deal with the first two arguments in this section. In the next section, I address Kass's third argument when I discuss Lucretius's argument that mortality is necessary for developing the virtues. I discussed Kass's fourth argument when I discussed Bernard Williams's argument that immortality is necessarily and intolerably boring.

3. See also President's Council on Bioethics, "Ageless Bodies," chap. 4 in *Beyond Therapy*; and Nicholas Agar, *Humanity's End: Why We Should Reject Radical Enhancement* (Cambridge, MA: MIT Press, 2010), 115–118. Aveek Bhattacharya and Robert Mark Simpson have questioned Agar's claim that extended life will be dominated by fear of death, partly on the ground that the odds of accidental death are low at any given time and people tend to ignore remote risks and partly on the ground that

even an extended life dominated by fear of death may well contain more net welfare than a nonextended life (and if it does not, one can always terminate life extension, resume aging, and die). "Life in Overabundance: Agar on Life-Extension and the Fear of Death," *Ethical Theory and Moral Practice* 17, no. 2 (2014): 223–236.

4. President's Council on Bioethics, "Ageless Bodies."

5. Hans Jonas, *The Imperative of Responsibility: In Search of an Ethics for the Technological Age* (Chicago: University of Chicago Press, 1985), 19; Kass, "L'Chaim and Its Limits," 313. See also Stephen G. Post, "Decelerated Aging: Should I Drink from a Fountain of Youth?," in *Fountain of Youth*, ed. Post and Binstock, 78.

6. Kass, "L'Chaim and Its Limits," 313.

7. John Pauley, "Agency, Identity and Technology: The Concealment of the Contingent in American Culture," *Janus Head* 10, no. 1 (2007): 43.

8. Post, "Decelerated Aging," 77.

9. Bill McKibben, *Enough: Staying Human in an Engineered Age* (New York: Henry Holt, 2003), 159.

10. Kass, "L'Chaim and Its Limits," 314.

11. Martha Nussbaum, *The Therapy of Desire* (Princeton: Princeton University Press, 1996), 227. Nussbaum makes these points in her discussion of Lucretius, who expressed his views about the desirability of immortality in a very long poem written in 50 BCE called "On the Nature of Things." In book 3 of that poem, he argues that the soul is mortal, then consoles the reader by arguing that there is such a thing as living too long:

Why is death so great a thing to you, mortal, that you give way excessively to sickly lamentation? Why groan and weep at death? For if the life that is past and gone has been pleasant to you, and all its blessings have not drained away and not been enjoyed—as if poured in a vessel full of holes—why don't you retire like a guest sated with the banquet of life, and with calm mind embrace, you fool, a rest that knows no care? (Translated by Cyril Bailey, http://www.humanistictexts.org/lucretius.htm [accessed on May 6, 2017])

I think this passage says there is a point when we have received as much benefit from life as it's possible to have, like becoming satiated at a feast, and that the value of the good things in life is subject to a law of diminishing returns. (That sounds like subjective meaning and boredom to me, but I'm not a classics scholar.) Nussbaum thinks Lucretius also meant something along the lines of her argument that some virtues cannot be developed in an immortal life, but I can't find that argument in Lucretius, so I attribute it to Nussbaum.

12. Nussbaum, 227–228.

13. Nussbaum, 228. Lisa Bortolotti has argued that these points can also be made in Kantian terms: Part of what's morally valuable about human beings is the fact that they can rationally choose their ends and act upon them. You can't develop the

ability to choose ends and act on them unless you learn to overcome obstacles, and this will be harder to do if you cannot die. "Agency, Life Extension, and the Meaning of Life," *Monist* 93, no. 1 (2010): 42–43.

14. This was noted by John Martin Fischer, "Contribution on Martha Nussbaum's *The Therapy of Desire*," *Philosophy and Phenomenological Research* 59, no. 3 (September 1999); and in Bortolotti, "Agency, Life Extension."

15. Fischer, "Contribution on Nussbaum," 791.

16. Bortolotti, "Agency, Life Extension," 45–49, 51.

17. Nussbaum, *The Therapy of Desire*, 228–229, 232.

18. Kass, "L'Chaim and Its Limits," 317.

19. Bruckner, "Defense of Adaptive Preferences," 311.

20. Maartje Schermer, "Preference Adaptation and Human Enhancement: Reflections on Autonomy and Well-Being," in *Adaptation and Autonomy: Adaptive Preferences in Enhancing and Ending Life*, ed. Juha Räikkä and Jukka Varelius (Berlin: Springer-Verlag, 2013), 119.

21. Jon Elster, *Sour Grapes: Studies in the Subversion of Rationality* (Cambridge: Cambridge University Press, 1985), 25.

22. Serene J. Khader, *Adaptive Preferences and Women's Empowerment* (Oxford: Oxford University Press, 2011), 116–119.

23. Jeff McMahan, *The Ethics of Killing: Problems at the Margins of Life* (Oxford: Oxford University Press, 2002), 140–141, 144.

Chapter 4

1. Portions of this chapter, particularly section 4.2, appear in another form in Davis, "Four Ways."

2. If declining life support counts as suicide, then surely stopping it by, for example, refusing further life-saving medications, counts as suicide too. Both actions intentionally shorten one's life.

3. Centers for Disease Control and Prevention, "Definitions: Self-Directed Violence," http://www.cdc.gov/violenceprevention/suicide/definitions.html (accessed on May 7, 2017).

4. *Oxford Dictionaries: American English*, s.v. "suicide," http://www.oxforddictionaries.com/us/definition/american_english/suicide (accessed on May 7, 2017).

5. Michael Cholbi, *Suicide: The Philosophical Dimensions* (London: Broadview Press, 2011), 21. I thank Michael Cholbi for providing helpful comments on a draft of this section, particularly those concerning moral injury and those concerning the killing/letting-die distinction.

6. Cholbi, 31.

7. Cholbi, 22, 31.

8. Franklin G. Miller, Robert D. Truog, and Dan W. Brock, "Moral Fictions and Medical Ethics," *Bioethics* 24, no. 9 (2010): 455. See also Dan W. Brock, "Life-Sustaining Treatment and Euthanasia," in *Encyclopedia of Bioethics*, ed. Stephen G. Post, 3rd ed. (New York: Macmillan Reference USA, 2004), 1418.

9. David Shaw, "The Body as Unwarranted Life Support: A New Perspective on Euthanasia," *Journal of Medical Ethics* 33, no. 9 (Spring 2007): 519–521.

10. This list comes from Cholbi, "The Moral Impermissibility of Suicide," chap. 2 in *Suicide*. To save space I've left out arguments that suicide is permissible. If you're interested in those arguments, see Cholbi, "The Moral Permissibility of Suicide," chap. 3 in *Suicide*.

11. Brett T. Litz et al., "Moral Injury and Moral Repair in War Veterans: A Preliminary Model and Intervention Strategy," *Clinical Psychology Review* 29 (2009): 697.

12. Shira Maguen and Brett Litz, "Moral Injury in Veterans of War," *PTSD Quarterly* 23, no. 1 (2012): 1.

Chapter 5

1. This section assembles a representative range of such worries. You can find such concerns expressed by a variety of writers, including the President's Council on Bioethics, "Ageless Bodies"; Francis Fukuyama, *Our Posthuman Future: Consequences of the Biotechnology Revolution* (New York: Picador Press, 2003), 57–71; and Leon R. Kass, "Ageless Bodies, Happy Souls: Biotechnology and the Pursuit of Perfection," *New Atlantis*, Spring 2003, 9–28, http://www.thenewatlantis.com/publications/ageless-bodies-happy-souls (accessed on May 7, 2017).

2. The Population Division of the UN Department of Economic and Social Affairs, *World Population Prospects: The 2015 Revision*, https://esa.un.org/unpd/wpp/publications/files/key_findings_wpp_2015.pdf (accessed on May 7, 2017).

3. Patrick Gerland et al., "World Population Stabilization Unlikely This Century," *Science* 346, no. 6206 (October 10, 2014): 234–237, https://doi.org/10.1126/science.1257469.

Chapter 6

1. Peter Singer, "Research into Aging: Should It Be Guided by the Interests of Present Individuals, Future Individuals, or the Species?," in *Life Span Extension: Consequences and Open Questions*, ed. Frederic C. Ludwig (New York: Springer, 1991), 138–139. See also Walter Glannon, "Extending the Human Life Span," *Journal of Medicine and Philosophy* 27, no. 30 (2002): 339–354; John Harris, "Intimations of Immortality," *Science* 288, no. 5463 (2000): 59; Daniel Kevles, "Life on the Far Side

of 150," *New York Times,* March 16, 1999, 27; Felicia Nimue Ackerman, "Death Is a Punch in the Jaw," in *The Oxford Handbook of Bioethics,* ed. Bonnie Steinbock (Oxford: Oxford University Press, 2007), 334.

2. Miller, "Extending Life," 170.

3. Gina Kolata, "Pushing Limits of the Human Life Span," *New York Times,* March 9, 1999.

4. *World Population Prospects.*

5. Global Footprint Network, "World Footprint: Do We Fit on the Planet?," http://www.footprintnetwork.org/en/index.php/GFN/page/world_footprint/ (accessed on May 7, 2017). See also The World Wildlife Fund, "How Many Planets Do You Need?," http://www.wwf.org.uk/about_wwf/press_centre/?unewsid=6783 (accessed on May 7, 2017).

6. This figure was extrapolated from the statistic that 95 percent of the developing world's population lives on less than $10 a day and does not include people in developed countries living at that level. See Martin Ravallion, Shaohua Chen, and Prem Sangraula, "Dollar a Day Revisited," *World Bank 2008,* http://elibrary .worldbank.org/doi/pdf/10.1596/1813-9450-4620 (accessed on May 7, 2017).

7. Jeff Wise, "About That Overpopulation Problem," *Slate: Future Tense,* January 9, 2013, http://www.slate.com/articles/technology/future_tense/2013/01/world_population _may_actually_start_declining_not_exploding.html (accessed on May 7, 2017).

8. *World Population Prospects.*

9. *World Population Prospects.*

10. This is a smaller increase than the increase for a population with a life expectancy of 150 years who has children at ages 25, 50, and 75, even though that population lives only 15 percent as long and has just as many children (table 2, appendix C). The main reason for this is that in the 1,000-year population, the third child is delayed until age 500. Spacing the births out that far dramatically slows the population growth, but not by enough, of course. Needless to say, if the 1,000-year population had the third child by age 500, the Malthusian catastrophe would materialize far, far sooner.

11. Although our tables and graphs were, for reasons of demographic method, based on the number of children per *woman,* the same age limit—as I said above—would apply to both men and women (and any other genders). Thus, if one member of a couple wins the right to reproduce at, say, age 50, it does not matter what age the other person is so far as keeping the population in check is concerned. The issue is not how late in a woman's life a child is born. The issue is how late in an average parent's life a child is born.

12. A DNA chip, also known as a DNA microarray, is a collection of microscopic DNA spots on a surface; put a drop of blood on the surface and the device will read the blood donor's genome.

Chapter 7

1. Portions of this chapter also appear in Davis, "Four Ways."

2. My case is based on two assumptions. First, I'm assuming that natural endowments, such as intelligence or genetic traits, are not subject to the requirements of distributive justice. Second, I'm assuming that comparative disadvantages we can't change, and which no one caused to happen, are also not subject to the requirements of distributive justice. (The second assumption might be the more fundamental one here.) Those assumptions are somewhat controversial. If you reject them, then you may reject my case, but my argument does not depend on those assumptions or that case, and the case is merely here to make it easier to think about distress apart from issues of justice.

3. There are "comparative theories of harm," but despite the term "comparative," comparative theories do not count being worse off than *someone else* as a harm. According to comparative theories, something harms you if it makes you worse off than *you* would otherwise have been. See Joel Feinberg, *Harmless Wrongdoing: The Moral Limits of the Criminal Law* (New York: Oxford University Press, 1988), 26.

4. See McMahan, "Death," chap. 2 in *The Ethics of Killing*. The final version of McMahan's account, leaving aside some details and qualifications, is this: how bad your death is turns on your overall losses from a variety of factors, including the particular way you died, where the loss consists of your best possible future with the most complete possible absence of the condition or event that caused your death and provided that future is realistic but discounted by your previous gains in life. McMahan, *The Ethics of Killing*, 183–184. Other versions of the counterfactual account appear in Thomas Nagel, "Death," in *Mortal Questions*, ed. Thomas Nagel (Cambridge: Cambridge University Press, 1991), 9; Joel Feinberg, *Harm to Others: The Moral Limits of the Criminal Law* (New York: Oxford University Press, 1984), 93; and Fred Feldman, *Confrontations with the Reaper: A Study of the Nature and Value of Death* (New York: Oxford University Press, 1994), 144.

5. Stephan Blatti, "Death's Distinctive Harm," *American Philosophical Quarterly* 49, no. 4 (2012): 323–324.

6. This objection assumes that society really can't afford to make life extension available to everyone, not just that it fails to do so. As I discuss in the next chapter, failure to make people better off in a way that preserves equality may count as a harm in a way that letting some people become better off when we cannot make everyone better off does not.

Chapter 8

1. Leonard Hayflick, "Modulating Aging, Longevity Determination and the Diseases of Old Age," in *Modulating Aging and Longevity*, ed. Suresh I. S. Rattan (Dordrecht, England: Kluwer Academic, 2003), 1–15.

2. Eric T. Juengst et al., "Aging: Antiaging Research and the Need for Public Dialogue," *Science* 299, no. 5611 (February 28, 2003): 1323.

3. Harris, "Intimations of Immortality," 59. See also John Harris, "A Response to Walter Glannon," *Bioethics* 16, no. 3 (2002): 290; Walter Glannon, "Reply to Harris," *Bioethics* 16, no. 3 (2002): 292–297.

4. Kass, "L'Chaim and Its Limits," 19.

5. One can also argue that the nightmare scenario is unjust because the wealthier members of society failed to perform their duty of beneficence toward the less well-off. Allen Buchanan has argued that because individuals have a duty of beneficence—a duty to increase or preserve the welfare or well-being of others—society should enforce that duty by taxing people to subsidize healthcare for the poor. "The Right to a Decent Minimum of Health Care," in *Justice and Health Care: Selected Essays*, ed. Allen Buchanan (Oxford: Oxford University Press, 2009), 28–36; Allen Buchanan, "Health-Care Delivery and Resource Allocation," in *Justice and Health Care*, ed. Buchanan, 71–75. Colin Farrelly has argued that the principle of beneficence requires developing life extension, though the details of his argument differ slightly from Buchanan's. "'Mind the Gap': Beneficence and Senescence," *Public Affairs Quarterly* 24, no. 2 (April 2010): 115–130.

Buchanan might be right about healthcare generally, but I don't think the duty of beneficence is a good basis for explaining what is unjust about the nightmare scenario. First, the duty of beneficence is usually understood as limited to alleviating or preventing harm, and as I argue in 10.4, not getting life extension is not a harm; it's a missing improvement in your condition. Second, most philosophers believe the duty of beneficence is weighted; that is, we have a duty to benefit others only when the cost to us is rather low compared to the amount of welfare conferred on others. The cost of providing life extension is likely to be quite high, so it is unlikely this conception of the duty of beneficence requires much of the Haves in the nightmare scenario. There are more demanding conceptions of the duty of beneficence, but they either require less than equality requires or require no more than equality requires, so they are otiose to our analysis.

6. Sophisticated, nuanced versions of utilitarianism and libertarianism will have the same implications as egalitarian theories in the nightmare scenario. Utilitarians must consider the diminishing value of additional goods, the disutility of insecurity, the utility of feelings of social solidarity, and other factors, which tend to have an equalizing effect. Nozick's libertarianism includes a principle of rectification for past injustices, and this principle has the potential to require greater equality, depending on how far society is from the distribution that would result if all original acquisitions were just and all transfers were voluntary, free of fraud, and so on.

7. Norman Daniels has argued that society should subsidize healthcare for the needy so that everyone has an equal opportunity to work and do other things that require good health. *Just Health Care: Meeting Health Needs Fairly* (Cambridge: Cambridge University Press, 2008), 63–77. Colin Farrelly has applied this argument to

life extension, contending that life extension is necessary for normal human functioning because it reduces or delays age-related diseases. "Equality and the Duty to Retard Human Aging," *Bioethics* 24, no. 8 (2010): 1–8. His egalitarian arguments differ from mine in this way: he contends that we have an egalitarian right to healthcare and then argues that life extension is part of the healthcare we have a right to.

8. Harry G. Frankfurt, "Equality as a Moral Ideal," in *Equality: Selected Readings*, ed. Louis P. Pojman and Robert Westmoreland (Oxford: Oxford University Press, 1997), 261–262.

9. Frankfurt, 271.

10. Frankfurt, 272, 273.

11. Glannon, "Reply to Harris," 292–293.

12. Joseph Raz, *The Morality of Freedom* (Oxford: Oxford University Press, 1986), 227.

13. John Rawls, *The Law of Peoples* (Cambridge, MA: Harvard University Press, 1999), 4–6, 89–91.

14. Colin Farrelly rejects the prior needs objection for a different reason. He argues that there is no morally relevant distinction between saving a life and extending a life. After all, we extend every life we save, and we save every life we extend. Therefore, we should not give priority to curing conditions that end life early over life extension. "Aging Research: Priorities and Aggregation," *Public Health Ethics* 1, no. 3 (2008): 258–267, 259–261.

I think he's right about this, though I think the prior needs objection doesn't rest on the claim that saving a life is more important than extending a life. The more plausible version of the objection is that those who die young have greater needs than those who enjoy a normal life span and wish to extend it. That too is debatable, of course. Curing a disease might save the patient 50 years of life, but extending life might save (or extend) life for a much longer span, and the greater gain might justify giving priority to extending life. However, I don't rely on that argument.

15. For example, the price of a console television set immediately after the Second World War started at $499 for a 10-inch screen and went as high as $2,495 for a 20-inch model. If these sets were bought in 1949, then adjusted for inflation, the cheaper set would now cost $4,825.37 and the expensive one $24,126.85—and that's not even in color. However, prices soon plummeted, and by the late 1960s, a single person on modest wages in a developed country could afford a color TV. "Television History: The First 75 Years," http://www.tvhistory.tv/tv-prices.htm (accessed on May 7, 2017).

Chapter 9

1. I'm not assuming that all welfare is a function of the amount of welfare a person has at a given time. As David Velleman puts it, welfare isn't always additive in that way. J. David Velleman, "Well-Being and Time," in *The Metaphysics of Death*, ed. Fischer, 329–357. Consider, for example, two lives. The first life is characterized by rising accomplishment and pleasure, the second by declining accomplishment and pleasure. In the rising life, early struggle and privation gradually gave way to increasing success and satisfaction, and old age was a time to reflect on triumph. In the declining life, the person expended the same effort and enjoyed the same amount of success and satisfaction, but there is something better about the rising life: it has a better, more attractive structure. The nonadditive aspect of total welfare per life does not affect our analysis; welfare per life-year is a function of total welfare (additive or otherwise) divided by the number of years lived, so the extra welfare in the rising life produces a greater amount of welfare per life-year, even though, year by year, the rising life doesn't contain more welfare in any given year than some counterpart year in the declining life. Besides, this kind of difference between lives will average out over an entire population.

2. The concept of a quality-adjusted life-year, or QALY, was developed by healthcare economists to measure the degree to which a given impairment or condition reduces the value of a year of life. A year of good health is worth 1.0 QALY. If you suffer from severe arthritis and you are indifferent between 10 years with arthritis and 9 years without it, then a year of living with that particular arthritis is worth 0.9 QALYs to you. These results are generalizable through polls and surveys.

QALYs are controversial. The controversy is not so much over whether QALYs are a plausible measure of what they purport to measure but over whether maximizing QALYs is the right way to make social decisions about how to allocate healthcare resources. The objections are essentially the same objections often raised against utilitarianism: maximizing welfare may require violating rights. Maximizing QALYs may, for example, call for favoring patients whose quality of life (absent their ailment) is good over patients whose quality of life has always been poorer (such as paraplegics), or treating a larger number of less serious problems over a smaller number of more serious problems. Right now, I am considering welfare alone; I am not claiming that maximizing QALYs is all we need do.

3. Singer, "Research into Aging," 138–139.

4. "Television History."

5. Zachary Karabell, "If the World Is Getting Richer, Why Do So Many People Feel Poor?," *Reuters*, January 24, 2014, http://blogs.reuters.com/edgy-optimist/2014/01/24/if-the-world-is-getting-richer-why-do-so-many-people-feel-poor/ (accessed on May 7, 2017); Charles Kenny, "The Poor Are Getting . . . Richer," *Foreign Policy*, February 7, 2011, http://foreignpolicy.com/2011/02/07/the-poor-are-getting-richer/ (accessed on May 7, 2017).

6. Andrew Walker, "A Richer World . . . but for Whom?," *BBC News*, January 21, 2015, http://www.bbc.com/news/business-30821925 (accessed on May 7, 2017).

7. These figures are in 1990 Geary-Khamis dollars, a unit representing what one US dollar would buy in America in 1990. Groningen Growth and Development Centre, "Statistics on World Population, GDP and Per Capital GDP, 1–2008," http://www.ggdc.net/MADDISON/oriindex.htm (accessed on May 7, 2017).

8. Nicholas Kristof, "The Most Important Thing, and It's Almost a Secret," *New York Times*, October 1, 2015, http://www.nytimes.com/2015/10/01/opinion/nicholas-kristof-the-most-important-thing-and-its-almost-a-secret.html?rref=collection%2Fcolumn%2Fnicholas-kristof&contentCollection=opinion&action=click&module=NextIn Collection&region=Footer&pgtype=article&_r=0 (accessed on May 7, 2017).

9. World Bank, "Regional Aggregation Using 2011 PPP and $1.9/Day Poverty Line," http://iresearch.worldbank.org/PovcalNet/index.htm?1 (accessed on May 7, 2017).

10. For an excellent discussion of these issues, see Gustaf Arrhenius, "Life Extension versus Replacement," *Journal of Applied Philosophy* 25, no. 3 (2008): 211–227.

11. Singer, "Research into Aging," 139, 143–144.

12. I am not the only one to question Singer's argument on empirical grounds. Mark Walker argues that Singer is wrong about two of his empirical claims: there is empirical evidence that people actually get happier in their elderly phase (not less happy, as Singer claims), and those who choose to extend their lives will be, on average, happier than the average of those who don't (the unhappy ones won't want to live longer). Therefore, according to Walker, an extended world will be happier and have a greater aggregate utility. "Superlongevity and Utilitarianism," *Australasian Journal of Philosophy* 85, no. 4 (2007): 581–595.

Chapter 10

1. I'm not saying that the arguments for rights are based on the consequences of respecting them as a general practice. One can argue for rights that way, or one can argue for them in other, nonconsequentialist ways.

2. Not everyone calls this a right in the strict sense. This is sometimes expressed as a restriction on paternalism or as a principle of respect for autonomy. Utilitarians do not speak of rights, but Mill's harm principle effectively directs us to respect autonomy and functions in practice much like a right. John S. Mill, *On Liberty*, in *Utilitarianism, On Liberty, Considerations on Representative Government*, by John Stuart Mill, ed. Geraint Williams (Rutland, VT: Charles E. Tuttle, 1993), 78.

3. For noncomparative accounts of harm, see Seana Shiffrin, "Wrongful Life, Procreative Responsibility, and the Significance of Harm," *Legal Theory* 5 (1999): 117–148; Jeff McMahan, "Wrongful Life: Paradoxes in the Morality of Causing People to Exist," in *Rational Commitment and Social Justice: Essays for Gregory Kavka*, ed. Jules Coleman (Cambridge: Cambridge University Press, 1998).

4. Joel Feinberg's comparative view of harm includes this counterfactual requirement. *Harmless Wrongdoing*, 26.

Chapter 11

1. Ronald Dworkin, "Taking Rights Seriously," chap. 7 in *Taking Rights Seriously*
(Cambridge, MA: Harvard University Press, 1977).

2. John Rawls, *A Theory of Justice* (Cambridge, MA: Harvard University Press, 1971),
535; John Rawls, *Justice as Fairness: A Restatement* (Cambridge, MA: Harvard University Press, 2001), 42–43, 87–88; Frankfurt, "Equality as a Moral Ideal"; Richard J. Arneson, "Equality," in *A Companion to Contemporary Political Philosophy*, ed. Robert E.
Goodin, Philip Pettit, and Thomas Pogge (Oxford: Basil Blackwell, 1993); Louis P.
Pojman and Robert Westmoreland, "Introduction: The Nature and Value of Equality," in *Equality: Selected Readings*, ed. Pojman and Westmoreland, 5.

3. For a view along these lines, see Dworkin, "Taking Rights Seriously," 200. I am
focused on when the right against harm can be outweighed by welfare, but there are
other situations where a right might be outweighed by something. A right can (and
must) be violated when one right conflicts with another right. A right might also
have less weight when the interests protected by the right are not really at stake
(perhaps the right to free speech is stronger when it comes to political issues and
weaker when it comes to pornography).

4. That last part is also true in the version of a veil of ignorance Rawls uses: "No
generation knows its place among the generations." Rawls, *Justice as Fairness*, 160.

5. R. M. Hare, "What Is Wrong with Slavery?," *Philosophy and Public Affairs* 8, no. 2
(1979): 103–121.

6. Of course, you might think that someone who would not take the gamble should
not have their welfare cut in half—that her refusal is a sufficient veto to make this
unjust (that's how the rights of the minority work against the interests of the majority, after all). However, we can't assume that the 1 percent who end up as Have-nots
were all people who refused to take that gamble. The right way to think about it is
this: If some rational agents would take that gamble and others would not, then
assume the 1 percent is representative of the entire population. In other words, if
10 percent of rational agents would not take the gamble, then we can say that the
1 percent who end up as Have-nots are 90 percent inclined to take the gamble and
that being substantially inclined to take it is effectively a consent to that distribution.

7. Pew Research Center, "Living to 120 and Beyond."

8. Partridge et al., "Public Attitudes."

Chapter 12

1. Kass, *Life, Liberty*; Kass, "The Wisdom of Repugnance," *New Republic* 216, no. 22 (June 2, 1997): 17–26; Kass, "Ageless Bodies, Happy Souls"; Kass, "L'Chaim and Its Limits"; Michael J. Sandel, *The Case against Perfection: Ethics in the Age of Genetic Engineering* (Cambridge, MA: Harvard University Press, 2009); Fukuyama, *Our Posthuman Future*; Jürgen Habermas, *The Future of Human Nature* (Cambridge: Polity Press, 2003); McKibben, *Enough*.

2. *Beyond Therapy* is available at http://bioethics.georgetown.edu/pcbe/reports/beyondtherapy/ (accessed on May 7, 2017), and *Human Cloning and Human Dignity* is available at http://bioethics.georgetown.edu/pcbe/reports/cloningreport/ (accessed on May 7, 2017).

3. Nick Bostrom, "The Transhumanist FAQ," Humanity+, http://humanityplus.org/philosophy/transhumanist-faq/#answer_19 (accessed on May 17, 2017).

4. Kass, "Ageless Bodies, Happy Souls," 10–13.

5. Kass, 20.

6. Nick Bostrom and Julian Savulescu discuss this argument in "Human Enhancement Ethics: The State of the Debate," in *Human Enhancement*, ed. Julian Savulescu and Nick Bostrom (Oxford: Oxford University Press, 2009), 2.

7. Frances Kamm, "What Is and Is Not Wrong with Enhancement," in *Human Enhancement*, ed. Savulescu and Bostrom, 109.

8. John Harris, "Enhancements Are a Moral Obligation," in *Human Enhancement*, ed. Savulescu and Bostrom, 131.

9. United Nations General Assembly, "Rio Declaration on Environment and Regulation," June 3–14, 1992, http://www.un.org/documents/ga/conf151/aconf15126-1annex1.htm/ (accessed on May 7, 2017).

10. Available at several websites, including http://www.sehn.org/precaution.html (accessed on May 7, 2017).

11. For an excellent extended criticism of the precautionary principle, see John Harris and Soren Holm, "Extended Lifespan and the Paradox of Precaution," *Journal of Medicine and Philosophy* 27, no. 3 (2002): 355–368.

12. Allen Buchanan, *Beyond Humanity? The Ethics of Biomedical Enhancement* (Oxford: Oxford University Press, 2011), 201.

13. Harris and Holm, "Extended Lifespan," 356. Harris and Holm also point out that the precautionary principle doesn't make sense as an epistemic principle; it calls for giving evidence pointing in one direction more weight than evidence pointing in the other direction when there's no epistemic reason to think the evidence in one direction is more strongly justified. Moreover, as a rule of choice, the principle

prevents us from ever acquiring evidence about the safety of something, and some of those things might turn out to be safe (361–362).

14. Harris and Holm reject the claim that life extension is a case where it's more risky to do something novel (365–366).

15. Buchanan, *Beyond Humanity?*, 201–202.

16. Norman Daniels discusses an argument similar to this in "Can Anyone Really Be Talking about Ethically Modifying Human Nature?," in *Human Enhancement*, ed. Savulescu and Bostrom, 38.

17. Erik Parens, "Toward a More Fruitful Debate about Enhancement," in *Human Enhancement*, ed. Savulescu and Bostrom, 182.

18. Kass, "Ageless Bodies, Happy Souls," 22–23.

19. Carl Elliott, "The Tyranny of Happiness," in *Enhancing Human Traits: Ethical and Social Implications*, ed. Erik Parens (Washington, DC: Georgetown University Press, 1998), 181–182.

20. Sandel, *The Case against Perfection*, 26–27.

21. Sandel, 85–90.

22. Sandel, 96.

23. Sandel, 96–97.

24. Richard Norman, "Interfering with Nature," *Journal of Applied Philosophy* 13, no. 1 (1996): 1–11.

25. John Danaher, "Hyperagency and the Good Life—Does Extreme Enhancement Threaten Meaning?," *Neuroethics* 7 (2014): 227. Danaher discusses similar hyperagency arguments from four other philosophers.

26. Norman, "Interfering with Nature," 6.

27. Norman, 6, 8–10.

28. Norman, 9–10.

29. Norman, 4.

30. Norman, 5.

31. Fukuyama, *Our Posthuman Future*, 147–151, 171–172.

32. Fukuyama objects to life extension, but his objections to life extension are not based on human nature. They are based on predictions of bad social consequences when lots of people undergo life extension—too many retired people, a general loss of social vitality, and so on. These are not enhancement concerns, and in any event, we discussed concerns like these in chapters 5 and 6.

33. For an overview of these arguments, see Allen Buchanan et al., *From Chance to Choice: Genetics and Justice* (Cambridge: Cambridge University Press, 2000), 110–124. Buchanan et al. acknowledge a limited role for the treatment/enhancement distinction

in health insurance but conclude that the distinction does not fully map onto the distinction between obligatory and nonobligatory medical services.

34. Timothy F. Murphy argues that what matters is not whether aging is a disease but whether it is objectionable and should be prevented and cured. "A Cure for Aging?," *Journal of Medicine and Philosophy* 11 (1986): 237–255.

35. Hayflick, *How and Why We Age*, 48.

36. Leonard Hayflick, "Anti-aging Medicine: Hype, Hope, and Reality," *Generations: Journal of the American Society on Aging* 25, no. 4 (November 30, 2000): 20–26.

37. Miller, "Extending Life," 159.

38. David Gems, "Aging: To Treat, or Not to Treat?," *American Scientist* 99, no. 4 (July/August 2011): 3, http://www.americanscientist.org/issues/pub/aging-to-treat-or -not-to-treat (accessed on May 7, 2017).

39. Arthur L. Caplan, "The 'Unnaturalness' of Aging—a Sickness Unto Death?," in *Concepts of Health and Disease: Interdisciplinary Perspectives*, ed. Arthur L. Caplan, H. Tristram Engelhardt, and James J. McCartney (Reading, MA: Addison-Wesley, 1981).

40. Edward J. Masoro, "Are Age-Associated Diseases an Integral Part of Aging?," *Handbook of the Biology of Aging*, ed. Masoro and Austad, 47.

41. Masoro, 45–46.

42. Masoro, 48.

43. Masoro, 49.

44. Timiras, "Comparative and Differential Aging," 36.

45. Hayflick, *How and Why We Age*, 47.

46. Masoro, "Age-Associated Diseases," 45.

47. Hayflick, *How and Why We Age*, 45–47.

Appendix A

1. Masoro, "Age-Associated Diseases," 44–45.

2. Sometimes aging is defined in terms of deterioration. For example, Robert Arking defines aging as "the time-independent series of cumulative, progressive, intrinsic, and deleterious . . . functional and structural changes that usually begin to manifest themselves at reproductive maturity and eventually culminate in death" (*Biology of Aging: Observations and Principles*, 2nd ed. [Sunderland, MA: Sinauer Associates, 1998], 12). This may look like a third definition, but I consider it an abbreviated version of the risk of failure definition; the implication is that deterioration gets worse over time and eventually gets bad enough to cause death. The risk of failure definition gets at this idea more precisely.

3. Quoted in Tom Kirkwood, *Time of Our Lives: The Science of Human Aging* (Oxford: Oxford University Press, 1999), 35.

4. Austad, *Why We Age*, 10.

5. Miller, "Extending Life," 166. See also Timiras, "Comparative and Differential Aging."

6. The same phenomenon occurs in fruit flies and the *C. Elegans* worm. S. D. Pletcher and J. W. Curtsinger, "Mortality Plateaus and the Evolution of Senescence: Why Are Old-Age Mortality Rates So Low?," *Evolution* 52 (1998): 454–464.

7. David Gems, "Is More Life Always Better? The New Biology of Aging and the Meaning of Life," *Hastings Center Report* 33, no. 4 (2003): 31.

8. Austad, *Why We Age*, 98–103; Kirkwood, *Time of Our Lives*, 78–79.

9. Consistent with the mutational accumulation theory, different species have different phenotypic changes associated with aging. Horses, for example, often die of twisted intestines, which almost never happens to humans, while only humans, monkeys, guinea pigs, and a few other species get Alzheimer's disease.

10. Michael R. Rose, *The Long Tomorrow: How Advances in Evolutionary Biology Can Help Us Postpone Aging* (Oxford: Oxford University Press, 2005), 26–27; Austad, *Why We Age*, 103–104; Kirkwood, *Time of Our Lives*, 77–78.

11. The main difference between the mutational accumulation theory and the antagonistic pleiotropy theory is that the mutational accumulation theory says that genes with harmful effects later in life are passively accumulated, while the antagonistic pleiotropy theory says they are favored by natural selection because of their beneficial early life effects.

12. Kirkwood, *Time of Our Lives*, 65–77.

13. Hayflick, *How and Why We Age*, 236–239.

14. Valter D. Longo, Joshua Mitteldorf, and Vladimir P. Skulachev, "Programmed and Altruistic Ageing," *Nature Reviews Genetics* 6 (November 2005): 866–872, https://doi.org/10.1038/nrg1706; T. C. Goldsmith, "On the Programmed/Non-programmed Aging Controversy," *Biochemistry (Moscow)* 77, no. 7 (July 2012): 729–732. For a contrary view, see Tom Kirkwood and Simon Melov, "On the Programmed/Non-programmed Nature of Ageing within the Life History," *Current Biology* 21, no. 18 (September 2011): R701–R707.

Appendix B

1. Williams, "The Makropulos Case," 84. Williams seems doubtful that personal identity can survive that degree of change over time, but he entertains the possibility in his discussion.

2. Williams, 78.

3. Williams's argument here resembles the regress argument against psychological egoism. The psychological egoist says that what you want is to have your desires satisfied. However, that leads to a regress, for at some point you have to want

something other than "whatever you want," lest your wants be empty. You have to want X for some reason other than the mere fact that X is something you want.

4. Williams, 77–78.

5. Williams, 83.

6. Williams believes it's impossible for a future self with a radically different character and values to retain categorical desires from an earlier phase, for what you desire depends in part on your character.

7. There is at least one exception to this: If I desire that my genetic lineage persist into the future, then the existence of my future self is necessary and sufficient for fulfilling that desire even if he does not care about it (assuming he doesn't do something rash, like alter his DNA just to spite his earlier self).

8. Williams, "The Makropulos Case," 83–84.

9. Temkin, "Is Living Longer Living Better?," 200–201.

10. This list comes from Derek Parfit, who develops it for reasons unrelated to life extension and immortality. *Reasons and Persons* (Oxford: Oxford University Press, 1984), 499.

Bibliography

Ackerman, Felicia Nimue. "Death Is a Punch in the Jaw." In *The Oxford Handbook of Bioethics*, ed. Bonnie Steinbock, 324–348. Oxford: Oxford University Press, 2007.

Agar, Nicholas. *Humanity's End: Why We Should Reject Radical Enhancement*. Cambridge, MA: MIT Press, 2010.

Aho, Kevin. "Simmel on Acceleration, Boredom, and Extreme Aesthesia." *Journal for the Theory of Social Behavior* 37, no. 4 (2007): 447–462.

Alvarez, Allen, Lumberto Mendoza, and Peter Danielson. "Mixed Views about Radical Life Extension." *Etikki i praksis—Nordic Journal of Applied Ethics* 9, no. 1 (2015): 87–110.

Arking, Robert. *Biology of Aging: Observations and Principles*. 2nd ed. Sunderland, MA: Sinauer Associates, 1998.

Arneson, Richard J. "Equality." In *A Companion to Contemporary Political Philosophy*, ed. Robert E. Goodin, Philip Pettit, and Thomas Pogge, 489–507. Oxford: Basil Blackwell, 1993.

Arrhenius, Gustaf. "Life Extension versus Replacement." *Journal of Applied Philosophy* 25, no. 3 (2008): 211–227.

Austad, Stephen N. *Why We Age: What Science Is Discovering about the Body's Journey through Life*. New York: John Wiley & Sons, 1997.

Bargdill, R. "The Study of Life Boredom." *Journal of Phenomenological Psychology* 31 (2000): 188–219.

Bhattacharya, Aveek, and Robert Mark Simpson. "Life in Overabundance: Agar on Life-Extension and the Fear of Death." *Ethical Theory and Moral Practice* 17, no. 2 (2014): 223–236.

Blackburn, Elizabeth H., Elissa S. Epel, and Jue Lin. "Human Telomere Biology: A Contributory and Interactive Factor in Aging, Disease Risks, and Protection." *Science* 350, no. 6265 (December 4, 2015): 1193–1198.

Blatti, Stephan. "Death's Distinctive Harm." *American Philosophical Quarterly* 49, no. 4 (2012): 317–330.

Bortolotti, Lisa. "Agency, Life Extension, and the Meaning of Life." *Monist* 93, no. 1 (2010): 38–56.

Bortolotti, Lisa, and Yujin Nagasawa. "Immortality without Boredom." *Ratio* 22, no. 3 (September 2009): 261–277.

Bostrom, Nick, and Julian Savulescu. "Human Enhancement Ethics: The State of the Debate." In *Human Enhancement*, ed. Julian Savulescu and Nick Bostrom, 1–22. Oxford: Oxford University Press, 2009.

Bostrom, Nick, and Julian Savulescu. "The Transhumanist FAQ." Humanity+. http://humanityplus.org/philosophy/transhumanist-faq/#answer_19 (accessed on May 17, 2017).

Brock, Dan W. "Life-Sustaining Treatment and Euthanasia." In *Encyclopedia of Bioethics*, ed. Stephen G. Post, 3rd ed., 1410–1421. New York: Macmillan Reference USA, 2004.

Bruckner, Donald W. "In Defense of Adaptive Preferences." *Philosophical Studies* 142, no. 3 (2009): 307–324.

Buchanan, Allen. *Beyond Humanity? The Ethics of Biomedical Enhancement*. Oxford: Oxford University Press, 2011.

Buchanan, Allen. "Health-Care Delivery and Resource Allocation." In *Justice and Health Care: Selected Essays*, ed. Allen Buchanan, 37–76. Oxford: Oxford University Press, 2009.

Buchanan, Allen. "The Right to a Decent Minimum of Health Care." In *Justice and Health Care: Selected Essays*, ed. Allen Buchanan, 28–36. Oxford: Oxford University Press, 2009.

Buchanan, Allen, Dan W. Brock, Norman Daniels, and Daniel Wikler. *From Chance to Choice: Genetics and Justice*. Cambridge: Cambridge University Press, 2000.

Buck Institute. "Buck Institute for Research on Aging." http://www.buckinstitute.org/ (accessed on May 2, 2017).

Calico. http://www.calicolabs.com/ (accessed on May 2, 2017).

Callahan, Daniel. *False Hopes: Why America's Quest for Perfect Health Is a Recipe for Failure*. New York: Simon and Schuster, 1998.

Caplan, Arthur L. "The 'Unnaturalness' of Aging—a Sickness Unto Death?" In *Concepts of Health and Disease: Interdisciplinary Perspectives*, ed. Arthur L. Caplan, H. Tristram Engelhardt, and James J. McCartney, 725–737. Reading, MA: Addison-Wesley, 1981.

Centers for Disease Control and Prevention. "Definitions: Self-Directed Violence." http://www.cdc.gov/violenceprevention/suicide/definitions.html (accessed on May 7, 2017).

Centers for Disease Control and Prevention. "Leading Causes of Death." http://www.cdc.gov/nchs/fastats/leading-causes-of-death.htm (accessed on May 2, 2017).

Chappell, Timothy. "Infinity Goes Up On Trial: Must Immortality Be Meaningless?" *European Journal of Philosophy* 17, no. 1 (2009): 30–44.

Cholbi, Michael. *Suicide: The Philosophical Dimensions*. London: Broadview Press, 2011.

Clark, William R. *A Means to an End: The Biological Basis of Aging and Death*. Oxford: Oxford University Press, 2002.

Cohen, Jon. "Death-Defying Experiments: Pushing the Limits of Life Span in Animals Could Someday Help Lengthen Our Own." *Science* 350, no. 6265 (December 4, 2015): 1186–1187.

Crew, Bec. "Japanese Scientists Reverse Ageing in Human Cell Lines: They Made a 97-Year-Old Cell Line Behave as Good as New." *Science Alert*, May 28, 2015. http://www.sciencealert.com/japanese-scientists-reverse-ageing-in-human-cell-lines (accessed on May 2, 2017).

Danaher, John. "Hyperagency and the Good Life—Does Extreme Enhancement Threaten Meaning?" *Neuroethics* 7 (2014): 227–242.

Daniels, Norman. "Can Anyone Really Be Talking about Ethically Modifying Human Nature?" In *Human Enhancement*, ed. Julian Savulescu and Nick Bostrom, 25–42. Oxford: Oxford University Press, 2009.

Daniels, Norman. *Just Health Care: Meeting Health Needs Fairly*. Cambridge: Cambridge University Press, 2008.

Davis, John K. "Four Ways Life Extension Will Change Our Relationship with Death." *Bioethics* 30, no. 3 (2016): 165–172.

de Beauvoir, Simone. *All Men Are Mortal*. Translated by Leonard M. Friedman. New York: W. W. Norton, 1992.

Delaney, Brian M., and Lisa Walford. *The Longevity Diet: The Only Proven Way to Slow the Aging Process and Maintain Peak Vitality through Caloric Restriction*. 3rd ed. Boston: Da Capo Lifelong Books, 2010.

Dell'Amore, Christine. "Old Mice Made 'Young'—May Lead to Anti-aging Treatments." *National Geographic*, January 7, 2012. http://news.nationalgeographic.com/news/2012/01/120106-aging-mice-stem-cells-old-young-science-health/ (accessed on May 2, 2017).

Dragojlovic, Nick. "Canadians' Support for Radical Life Extension Resulting from Advances in Regenerative Medicine." *Journal of Aging Studies* 27, no. 2 (2013): 151–158.

Dworkin, Ronald. "Taking Rights Seriously." In *Taking Rights Seriously* chap. 7. Cambridge, MA: Harvard University Press, 1977.

Elliot, Natalie. "The Politics of Life Extension in Francis Bacon's *Wisdom of the Ancients*." *Review of Politics* 77 (2015): 351–375.

Elliott, Carl. "The Tyranny of Happiness." In *Enhancing Human Traits: Ethical and Social Implications*, ed. Erik Parens, 177–188. Washington, DC: Georgetown University Press, 1998.

Elster, Jon. *Sour Grapes: Studies in the Subversion of Rationality*. Cambridge: Cambridge University Press, 1985.

Farrelly, Colin. "Aging Research: Priorities and Aggregation." *Public Health Ethics* 1, no. 3 (2008): 258–267.

Farrelly, Colin. "Equality and the Duty to Retard Human Aging." *Bioethics* 24, no. 8 (2010): 1–8.

Farrelly, Colin. "'Mind the Gap': Beneficence and Senescence." *Public Affairs Quarterly* 24, no. 2 (April 2010): 115–130.

Feinberg, Joel. *Harmless Wrongdoing: The Moral Limits of the Criminal Law*. New York: Oxford University Press, 1988.

Feinberg, Joel. *Harm to Others: The Moral Limits of the Criminal Law*. New York: Oxford University Press, 1984.

Feldman, Fred. *Confrontations with the Reaper: A Study of the Nature and Value of Death*. New York: Oxford University Press, 1994.

Fischer, John Martin. "Contribution on Martha Nussbaum's *The Therapy of Desire*." *Philosophy and Phenomenological Research* 59, no. 3 (September 1999): 787–792.

Fischer, John Martin. "Immortality." In *The Oxford Handbook of Philosophy of Death*, ed. Ben Bradley, Fred Feldman, and Jens Johansson, 336–354. Oxford: Oxford University Press, 2012.

Frankfurt, Harry G. "Equality as a Moral Ideal." In *Equality: Selected Readings*, ed. Louis P. Pojman and Robert Westmoreland, 261–273. Oxford: Oxford University Press, 1997.

Friend, Tad. "Silicon Valley's Quest to Live Forever." *New Yorker*, April 3, 2017. http://www.newyorker.com/magazine/2017/04/03/silicon-valleys-quest-to-live -forever (accessed on May 5, 2017).

Fukada, So-ichiro, Yuran Ma, and Akiyoshi Uezumi. "Adult Stem Cell and Mesenchymal Progenitor Theories of Aging." *Frontiers in Cell and Developmental Biology* 2, no. 10 (March 28, 2014): 1–9.

Fukuyama, Francis. *Our Posthuman Future: Consequences of the Biotechnology Revolution*. New York: Picador Books, 2003.

Gelles, David. "Immortality 2.0: A Silicon Valley Insider Looks at California's Transhumanist Movement." *Futurist* 43, no. 1 (2009): 34–41.

Gems, David. "Aging: To Treat, or Not to Treat?" *American Scientist*, July/August 2011. http://www.americanscientist.org/issues/pub/aging-to-treat-or-not-to-treat (accessed on May 7, 2017).

Gems, David. "Is More Life Always Better? The New Biology of Aging and the Meaning of Life." *Hastings Center Report* 33, no. 4 (2003): 31–39.

Gerland, Patrick, et al. "World Population Stabilization Unlikely This Century." *Science* 346, no. 6206 (October 10, 2014): 234–237. https://doi.org/10.1126/science.1257469.

Glannon, Walter. "Extending the Human Life Span." *Journal of Medicine and Philosophy* 27, no. 30 (2002): 339–354.

Glannon, Walter. "Identity, Prudential Concern, and Extended Lives." *Bioethics* 16, no. 3 (2002): 266–283.

Glannon, Walter. "Reply to Harris." *Bioethics* 16, no. 3 (2002): 292–297.

Glenn Foundation for Medical Research. http://www.glennfoundation.org/ (accessed on May 2, 2017).

Global Footprint Network. "World Footprint: Do We Fit on the Planet?" http://www.footprintnetwork.org/en/index.php/GFN/page/world_footprint/ (accessed on May 7, 2017).

Goldsmith, T. C. "On the Programmed/Non-programmed Aging Controversy." *Biochemistry (Moscow)* 77, no. 7 (July 2012): 729–732.

Groningen Growth and Development Centre. "Statistics on World Population, GDP and Per Capital GDP, 1–2008." http://www.ggdc.net/MADDISON/oriindex.htm (accessed on May 7, 2017).

Habermas, Jürgen. *The Future of Human Nature.* Cambridge: Polity Press, 2003.

Hall, Stephen S. "Kenyon's Ageless Quest." *Smithsonian* (March 2004), http://www.smithsonianmag.com/science-nature/kenyons-ageless-quest-103525532/?all (accessed on May 2, 2017).

Hare, R. M. "What Is Wrong with Slavery?" *Philosophy and Public Affairs* 8, no. 2 (1979): 103–121.

Harris, John. "Enhancements Are a Moral Obligation." In *Human Enhancement*, ed. Julian Savulescu and Nick Bostrom, 131–154. Oxford: Oxford University Press, 2009.

Harris, John. "Intimations of Immortality." *Science* 288, no. 5463 (2000): 59.

Harris, John. "A Response to Walter Glannon." *Bioethics* 16, no. 3 (2002): 284–291.

Harris, John, and Soren Holm. "Extended Lifespan and the Paradox of Precaution." *Journal of Medicine and Philosophy* 27, no. 3 (2002): 355–368.

Hayflick, Leonard. "Anti-aging Medicine: Hype, Hope, and Reality." *Generations: Journal of the American Society on Aging* 25, no. 4 (November 30, 2000): 20–26.

Hayflick, Leonard. *How and Why We Age.* New York: Ballantine Books, 1994.

Hayflick, Leonard. "Modulating Aging, Longevity Determination and the Diseases of Old Age." In *Modulating Aging and Longevity*, ed. Suresh I. S. Rattan, 1–15. Dordrecht, England: Kluwer Academic, 2003.

Hekimi, Siegfried, and Leonard Guarente. "Genetics and the Specificity of the Aging Process." *Science* 299, no. 5611 (2003): 1351–1354.

Holden, Constance. "The Quest to Reverse Time's Toll." *Science* 295, no. 5557 (February 8, 2002): 1032–1033.

Jonas, Hans. *The Imperative of Responsibility: In Search of an Ethics for the Technological Age*. Chicago: University of Chicago Press, 1985.

Juengst, Eric T., Robert H. Binstock, Maxwell J. Mehlman, and Stephen G. Post. "Aging: Antiaging Research and the Need for Public Dialogue." *Science* 299, no. 5611 (February 28, 2003): 1323.

Kaeberlein, Matt, Peter S. Rabinovitch, and George M. Martin. "Healthy Aging: The Ultimate Preventative Medicine." *Science* 350, no. 6265 (December 4, 2015): 1191–1193.

Kamm, Frances. "What Is and Is Not Wrong with Enhancement." In *Human Enhancement*, ed. Julian Savulescu and Nick Bostrom, 91–130. Oxford: Oxford University Press, 2009.

Kass, Leon R. "Ageless Bodies, Happy Souls: Biotechnology and the Pursuit of Perfection." *New Atlantis*, Spring 2003, 9–28. http://www.thenewatlantis.com/publications/ageless-bodies-happy-souls (accessed on May 7, 2017).

Kass, Leon R. "The Case for Mortality." *American Scholar* 52, no. 2 (Spring 1983): 173–191.

Kass, Leon R. "L'Chaim and Its Limits: Why Not Immortality?" *First Things* 113 (May 2001): 17–24.

Kass, Leon R. "L'Chaim and Its Limits: Why Not Immortality?" In *The Fountain of Youth: Cultural, Scientific, and Ethical Perspectives on a Biomedical Goal*, ed. Stephen G. Post and Robert H. Binstock, 304–320. Oxford: Oxford University Press, 2004.

Kass, Leon R. *Life, Liberty and the Defense of Dignity: The Challenge for Bioethics*. New York: Encounter Books, 2002.

Kass, Leon R. "The Wisdom of Repugnance." *New Republic* 216, no. 22 (June 2, 1997): 17–26.

Kenny, Charles. "The Poor Are Getting . . . Richer." *Foreign Policy*, February 7, 2011. http://foreignpolicy.com/2011/02/07/the-poor-are-getting-richer/ (accessed on May 7, 2017).

Khader, Serene J. *Adaptive Preferences and Women's Empowerment*. Oxford: Oxford University Press, 2011.

Kirkwood, Tom. *Time of Our Lives: The Science of Human Aging*. Oxford: Oxford University Press, 1999.

Kirkwood, Tom, and Simon Melov. "On the Programmed/Non-programmed Nature of Ageing within the Life History." *Current Biology* 21, no. 18 (September 2011): R701–R707.

Lane, Mark A., Donald K. Ingram, and George S. Roth. "The Serious Search for an Anti-aging Pill." *Scientific American* 287, no. 2 (August 2002): 36–41.

Lawrence Ellison Foundation. http://www.ellisonfoundation.org/ (accessed on May 2, 2017).

Leslie, Mitch. "The Secret of Antiaging Is Worth Looking For." *Science* 375, no. 6350 (October 8, 2013): http://news.sciencemag.org/economics/2013/10/secret-antiaging -worth-looking (accessed on May 2, 2017).

Litz, Brett T., et al. "Moral Injury and Moral Repair in War Veterans: A Preliminary Model and Intervention Strategy." *Clinical Psychology Review* 29 (2009): 695–706.

Longo, Valter D., Joshua Mitteldorf, and Vladimir P. Skulachev. "Programmed and Altruistic Ageing." *Nature Reviews Genetics* 6 (November 2005): 866–872. https://doi .org/10.1038/nrg1706.

Lucretius. "On the Nature of Things." Translated by Cyril Bailey. http://www .humanistictexts.org/lucretius.htm (accessed on May 6, 2017).

Maguen, Shira, and Brett Litz. "Moral Injury in Veterans of War." *PTSD Quarterly* 23, no. 1 (2012): 1–3.

Martin, George M., Kelly LaMarco, Evelyn Strauss, and Katrina L. Kelner. "Research on Aging: The End of the Beginning." *Science* 299 (February 28, 2003): 1339–1341.

Masoro, Edward J. "Are Age-Associated Diseases an Integral Part of Aging?" In *Handbook of the Biology of Aging*, ed. Edward J. Masoro and Steven N. Austad, 6th ed., 43–62. London: Academic Press, 2006.

McKibben, Bill. *Enough: Staying Human in an Engineered Age*. New York: Henry Holt, 2003.

McMahan, Jeff. *The Ethics of Killing: Problems at the Margins of Life*. Oxford: Oxford University Press, 2002.

McMahan, Jeff. "Wrongful Life: Paradoxes in the Morality of Causing People to Exist." In *Rational Commitment and Social Justice: Essays for Gregory Kavka*, ed. Jules Coleman, 208–247. Cambridge: Cambridge University Press, 1998.

McNicoll, Arion. "How Google's Calico Aims to Fight Aging and 'Solve Death.'" *CNNTech* (October 3, 2013), http://www.cnn.com/2013/10/03/tech/innovation/ google-calico-aging-death/ (accessed on May 2, 2017).

Methuselah Foundation. https://mfoundation.org/ (accessed on May 2, 2017).

Mill, John S. *On Liberty*. In John Stuart Mill, *Utilitarianism, On Liberty, Considerations on Representative Government*, ed. Geraint Williams. Rutland, VT: Charles E. Tuttle, 1993.

Miller, Franklin G., Robert D. Truog, and Dan W. Brock. "Moral Fictions and Medical Ethics." *Bioethics* 24, no. 9 (2010): 453–460.

Miller, Richard. "Extending Life: Scientific Prospects and Political Obstacles." *Milbank Quarterly* 80, no. 1 (2002): 155–174.

Murphy, Timothy F. "A Cure for Aging?" *Journal of Medicine and Philosophy* 11 (1986): 237–255.

Nagel, Thomas. "Death." In *Mortal Questions*, ed. Thomas Nagel, 1–10. Cambridge: Cambridge University Press, 1991.

National Institute of Aging. "Fiscal Year 2017 Budget." https://www.nia.nih.gov/about/budget/2016/fy-2017-justification-budget-request/fy-2017-program-descriptions-and (accessed on May 2, 2017).

National Institutes of Health. "Estimates of Funding for Various Research, Condition, and Disease Categories (RCDC)." http://report.nih.gov/categorical_spending.aspx (accessed on May 2, 2017).

Norman, Richard. "Interfering with Nature." *Journal of Applied Philosophy* 13, no. 1 (1996): 1–11.

Nozick, Robert. *Anarchy, State, and Utopia*. New York: Basic Books, 1974.

Nussbaum, Martha. *The Therapy of Desire*. Princeton: Princeton University Press, 1996.

Olshansky, S. Jay. "Replacement Parts: A Survey of Recent Articles." *Wilson Quarterly* 26 (Summer 2002): 101–103.

Olshansky, S. Jay, Daniel Perry, Richard A. Miller, and Robert N. Butler. "In Pursuit of the Longevity Dividend." *Scientist* (March 1, 2006): 28–36. http://www.the-scientist.com/?articles.view/articleNo/23784/title/The-Longevity-Dividend/ (accessed on May 2, 2017).

Olshansky, S. Jay, et al. "Position Statement on Human Aging." *Journals of Gerontology Series A: Biological Sciences and Medical Sciences* 57 (2002): B292–B297.

Overall, Christine. *Aging, Death, and Human Longevity: A Philosophical Inquiry*. Berkeley: University of California Press, 2003.

Parens, Erik. "Toward a More Fruitful Debate about Enhancement." In *Human Enhancement*, ed. Julian Savulescu and Nick Bostrom, 181–197. Oxford: Oxford University Press, 2009.

Parfit, Derek. *Reasons and Persons*. Oxford: Oxford University Press, 1984.

Park, Alice. "Age Disrupters: A Drug from Dirt and Some Siamese Mice Have Researchers Inching toward the Seemingly Impossible: A Cure for Aging." *Time* 185, nos. 6–7 (February 23, 2015–March 2, 2015): 73–76.

Partridge, Brad, Jayne Lucke, Helen Bartlett, and Wayne Hall. "Public Attitudes towards Human Life Extension by Intervening in Ageing." *Journal of Aging Studies* 25, no. 2 (2011): 73–83.

Partridge, Brad, Malr Underwood, Jayne Lucke, Helen Bartlett, and Wayne Hall. "Ethical Concerns in the Community about Technologies to Extend Human Life Span." *American Journal of Bioethics* 9, no. 12 (2009): 68–76.

Pauley, John. "Agency, Identity and Technology: The Concealment of the Contingent in American Culture." *Janus Head* 10, no. 1 (2007): 41–61.

Peng, Y., M. Xuan, V. Y. Leung, and B. Cheng. "Stem Cells and Aberrant Signaling of Molecular Systems in Skin Aging." *Ageing Research Review* 19 (January 2015): 8–21.

Pew Research Center. "Living to 120 and Beyond." http://www.pewforum.org/2013/08/06/living-to-120-and-beyond/ (accessed on May 2, 2017).

Pletcher, S. D., and J. W. Curtsinger. "Mortality Plateaus and the Evolution of Senescence: Why Are Old-Age Mortality Rates So Low?" *Evolution* 52 (1998): 454–464.

Pojman, Louis P., and Robert Westmoreland. "Introduction: The Nature and Value of Equality." In *Equality: Selected Readings*, ed. Louis P. Pojman and Robert Westmoreland, 1–14. Oxford: Oxford University Press, 1997.

Population Division of the UN Department of Economic and Social Affairs. *World Population Prospects: The 2015 Revision*. https://esa.un.org/unpd/wpp/publications/files/key_findings_wpp_2015.pdf (accessed on May 7, 2017).

Post, Stephen G. "Decelerated Aging: Should I Drink from a Fountain of Youth?" In *The Fountain of Youth: Cultural, Scientific, and Ethical Perspectives on a Biomedical Goal*, ed. Stephen G. Post and Robert H. Binstock, 72–93. Oxford: Oxford University Press, 2004.

President's Council on Bioethics. *Beyond Therapy*. http://bioethics.georgetown.edu/pcbe/reports/beyondtherapy/ (accessed on May 6, 2017).

President's Council on Bioethics. *Human Cloning and Human Dignity*. http://bioethics.georgetown.edu/pcbe/reports/cloningreport/ (accessed on May 7, 2017).

Promislow, Daniel E. L., Kenneth M. Fedorka, and Joep M. S. Burger. "Evolutionary Biology of Aging: Future Directions." In *Handbook of the Biology of Aging*, ed. Edward J. Masoro and Steven N. Austad, 6th ed., 217–242. London: Academic Press, 2006.

Ravallion, Martin, Shaohua Chen, and Prem Sangraula. "Dollar a Day Revisited." *World Bank 2008*. http://elibrary.worldbank.org/doi/pdf/10.1596/1813-9450-4620 (accessed on May 7, 2017).

Rawls, John. *Justice as Fairness: A Restatement*. Cambridge, MA: Harvard University Press, 2001.

Rawls, John. *The Law of Peoples*. Cambridge, MA: Harvard University Press, 1999.

Rawls, John. *A Theory of Justice*. Cambridge, MA: Harvard University Press, 1971.

Raz, Joseph. *The Morality of Freedom*. Oxford: Oxford University Press, 1986.

Richel, Theo. "Will Human Life Expectancy Quadruple in the Next Hundred Years? Sixty Gerontologists Say Public Debate of Life Extension Is Necessary." *Journal of Anti-aging Medicine* 6, no. 4 (2003): 309–314.

Rose, Michael R. *The Long Tomorrow: How Advances in Evolutionary Biology Can Help Us Postpone Aging*. Oxford: Oxford University Press, 2005.

Sandel, Michael J. *The Case against Perfection: Ethics in the Age of Genetic Engineering*. Cambridge, MA: Harvard University Press, 2009.

Schermer, Maartje. "Preference Adaptation and Human Enhancement: Reflections on Autonomy and Well-Being." In *Adaptation and Autonomy: Adaptive Preferences in*

Enhancing and Ending Life, ed. Juha Räikkä and Jukka Varelius, 117–136. Berlin: Springer-Verlag, 2013.

Schloendorn, John. "Making the Case for Human Life Extension: Personal Arguments." *Bioethics* 20, no. 4 (2006): 191–202.

SENS Research Foundation. "Annual Report." http://www.sens.org/sites/srf.org/files/ reports/SENS%20Research%20Foundation%20Annual%20Report%202015.pdf (accessed on May 2, 2017).

Shaw, David. "The Body as Unwarranted Life Support: A New Perspective on Euthanasia." *Journal of Medical Ethics* 33, no. 9 (Spring 2007): 519–521.

Shiffrin, Seana. "Wrongful Life, Procreative Responsibility, and the Significance of Harm." *Legal Theory* 5 (1999): 117–148.

Singer, Peter. "Research into Aging: Should It Be Guided by the Interests of Present Individuals, Future Individuals, or the Species?" In *Life Span Extension: Consequences and Open Questions*, ed. Frederic C. Ludwig, 132–145. New York: Springer, 1991.

Sousounis, Konstantinos, Joelle A. Baddour, and Panagiotis A. Tsonis. "Aging and Regeneration in Vertebrates." *Current Topics in Developmental Biology* 108 (2014): 217–246.

Speakman, John R., and Sharon E. Mitchell. "Caloric Restriction." *Molecular Aspects of Medicine* 32, no. 3 (2011): 159–221. http://www.sciencedirect.com/science/article/ pii/S009829971100032X (accessed on May 6, 2017).

Stein, Phyllis K., et al. "Caloric Restriction May Reverse Age-Related Autonomic Decline in Humans." *Aging Cell* 11, no. 4 (May 21, 2012): 644–650. http://onlinelibrary .wiley.com/doi/10.1111/j.1474-9726.2012.00825.x/full (accessed on May 7, 2017).

Stipp, David. "Beyond Resveratrol: The Anti-aging NAD Fad." *Scientific American Guest Blog*, March 11, 2015. https://blogs.scientificamerican.com/guest-blog/beyond -resveratrol-the-anti-aging-nad-fad/ (accessed on May 6, 2017).

Stipp, David. "The Hunt for the Youth Pill." *Fortune* 140, no. 7 (1999): 199–216.

Sutphin, George L., and Matt Kaeberlein. "Comparative Genetics of Aging." In *Handbook of the Biology of Aging*, ed. Edward J. Masoro and Steven N. Austad, 7th ed., 215–241. London: Academic Press, 2011.

Temkin, Larry. "Is Living Longer Living Better?" *Journal of Applied Philosophy* 25, no. 3 (August 2008): 193–210.

Timiras, Paola S. "Comparative and Differential Aging, Geriatric Functional Assessment, Aging and Disease." In *Physiological Basis of Aging and Geriatrics*, ed. Paola S. Timiras, 3rd ed., 25–46. New York: CRC Press, 2003.

Underwood, Emily. "The Final Countdown: In the Race to Find a Biological Clock, There Are Plenty of Contenders." *Science* 350, no. 6265 (December 4, 2015): 1188–1190.

United Nations General Assembly. "Rio Declaration on Environment and Regulation." 1992. http://www.un.org/documents/ga/conf151/aconf15126-1annex1.htm/ (accessed on May 7, 2017).

Velleman, J. David. "Well-Being and Time." In *The Metaphysics of Death*, ed. John Martin Fischer, 329–357. Stanford: Stanford University Press, 1993.

Verdin, Eric. "NAD in Aging, Metabolism, and Neurodegeneration." *Science* 350, no. 6265 (December 4, 2015): 1208–1213.

Walker, Mark. "Superlongevity and Utilitarianism." *Australasian Journal of Philosophy* 85, no. 4 (2007): 581–595.

Wang, Ying, and Siegfried Hekimi. "Mitochondrial Dysfunction and Longevity in Animals: Untangling the Knot." *Science* 350, no. 6265 (December 4, 2015): 1204–1207.

Weindruch, Richard. "Caloric Restriction and Aging." *Scientific American* 274, no. 1 (1996): 46–52.

Williams, Bernard. "The Makropulos Case: Reflections on the Tedium of Immortality." In *The Metaphysics of Death*, ed. John Martin Fischer, 73–92. Stanford: Stanford University Press, 1993.

Wingspread Conference on the Precautionary Principle. http://www.sehn.org/precaution.html (accessed on May 7, 2017).

Wise, Jeff. "About That Overpopulation Problem." *Slate: Future Tense*, January 9, 2013. http://www.slate.com/articles/technology/future_tense/2013/01/world_population_may_actually_start_declining_not_exploding.html (accessed on May 7, 2017).

World Bank. "Regional Aggregation Using 2011 PPP and $1.9/Day Poverty Line." http://iresearch.worldbank.org/PovcalNet/index.htm?1 (accessed on May 7, 2017).

World Wildlife Fund. "How Many Planets Do You Need?" http://www.wwf.org.uk/about_wwf/press_centre/?unewsid=6783 (accessed on May 7, 2017).

Wright, Karen, and Mary Ellen Mark. "Staying Alive." *Discover*, November 6, 2003. http://discovermagazine.com/2003/nov/cover (accessed on May 2, 2017).

Yirka, Bob. "Researchers Find a Link between Age Related Memory Loss and a Specific Enzyme in Mice." *Medical Press*, July 2, 2012. http://medicalxpress.com/news/2012-07-link-age-memory-loss-specific.html (accessed on May 2, 2017).

Index

Basic Bioethics

Arthur Caplan, editor

Books Acquired under the Editorship of Glenn McGee and Arthur Caplan

Peter A. Ubel, *Pricing Life: Why It's Time for Health Care Rationing*

Mark G. Kuczewski and Ronald Polansky, eds., *Bioethics: Ancient Themes in Contemporary Issues*

Suzanne Holland, Karen Lebacqz, and Laurie Zoloth, eds., *The Human Embryonic Stem Cell Debate: Science, Ethics, and Public Policy*

Gita Sen, Asha George, and Piroska Östlin, eds., *Engendering International Health: The Challenge of Equity*

Carolyn McLeod, *Self-Trust and Reproductive Autonomy*

Lenny Moss, *What Genes Can't Do*

Jonathan D. Moreno, ed., *In the Wake of Terror: Medicine and Morality in a Time of Crisis*

Glenn McGee, ed., *Pragmatic Bioethics*, 2nd edition

Timothy F. Murphy, *Case Studies in Biomedical Research Ethics*

Mark A. Rothstein, ed., *Genetics and Life Insurance: Medical Underwriting and Social Policy*

Kenneth A. Richman, *Ethics and the Metaphysics of Medicine: Reflections on Health and Beneficence*

David Lazer, ed., *DNA and the Criminal Justice System: The Technology of Justice*

Harold W. Baillie and Timothy K. Casey, eds., *Is Human Nature Obsolete? Genetics, Bioengineering, and the Future of the Human Condition*

Robert H. Blank and Janna C. Merrick, eds., *End-of-Life Decision Making: A Cross-National Study*

Norman L. Cantor, *Making Medical Decisions for the Profoundly Mentally Disabled*

Margrit Shildrick and Roxanne Mykitiuk, eds., *Ethics of the Body: Post-conventional Challenges*

Alfred I. Tauber, *Patient Autonomy and the Ethics of Responsibility*

David H. Brendel, *Healing Psychiatry: Bridging the Science/Humanism Divide*

Jonathan Baron, *Against Bioethics*

Michael L. Gross, *Bioethics and Armed Conflict: Moral Dilemmas of Medicine and War*

Karen F. Greif and Jon F. Merz, *Current Controversies in the Biological Sciences: Case Studies of Policy Challenges from New Technologies*

Deborah Blizzard, *Looking Within: A Sociocultural Examination of Fetoscopy*

Ronald Cole-Turner, ed., *Design and Destiny: Jewish and Christian Perspectives on Human Germline Modification*

Holly Fernandez Lynch, *Conflicts of Conscience in Health Care: An Institutional Compromise*

Mark A. Bedau and Emily C. Parke, eds., *The Ethics of Protocells: Moral and Social Implications of Creating Life in the Laboratory*

Jonathan D. Moreno and Sam Berger, eds., *Progress in Bioethics: Science, Policy, and Politics*

Eric Racine, *Pragmatic Neuroethics: Improving Understanding and Treatment of the Mind-Brain*

Martha J. Farah, ed., *Neuroethics: An Introduction with Readings*

Jeremy R. Garrett, ed., *The Ethics of Animal Research: Exploring the Controversy*

Books Acquired under the Editorship of Arthur Caplan

Sheila Jasanoff, ed., *Reframing Rights: Bioconstitutionalism in the Genetic Age*

Christine Overall, *Why Have Children? The Ethical Debate*

Yechiel Michael Barilan, *Human Dignity, Human Rights, and Responsibility: The New Language of Global Bioethics and Bio-law*

Tom Koch, *Thieves of Virtue: When Bioethics Stole Medicine*

Timothy F. Murphy, *Ethics, Sexual Orientation, and Choices about Children*

Daniel Callahan, *In Search of the Good: A Life in Bioethics*

Robert Blank, *Intervention in the Brain: Politics, Policy, and Ethics*

Gregory E. Kaebnick and Thomas H. Murray, eds., *Synthetic Biology and Morality: Artificial Life and the Bounds of Nature*

Dominic A. Sisti, Arthur L. Caplan, and Hila Rimon-Greenspan, eds., *Applied Ethics in Mental Healthcare: An Interdisciplinary Reader*

Barbara K. Redman, *Research Misconduct Policy in Biomedicine: Beyond the Bad-Apple Approach*

Russell Blackford, *Humanity Enhanced: Genetic Choice and the Challenge for Liberal Democracies*

Nicholas Agar, *Truly Human Enhancement: A Philosophical Defense of Limits*

Bruno Perreau, *The Politics of Adoption: Gender and the Making of French Citizenship*

Carl Schneider, *The Censor's Hand: The Misregulation of Human-Subject Research*

Lydia S. Dugdale, ed., *Dying in the Twenty-First Century: Towards a New Ethical Framework for the Art of Dying Well*

John D. Lantos and Diane S. Lauderdale, *Preterm Babies, Fetal Patients, and Childbearing Choices*

Harris Wiseman, *The Myth of the Moral Brain*

Arthur L. Caplan and Jason Schwartz, eds., *Vaccine Ethics and Policy: An Introduction with Readings*

Tom Koch, *Ethics in Everyday Places: Mapping Moral Stress, Distress, and Injury*

Nicole Piemonte, *Afflicted: How Vulnerability Can Heal Medical Education and Practice*

Printed in the United States
by Baker & Taylor Publisher Services